FARMING STORIES FROM
THE SCOTTISH BORDERS

FARMING STORIES FROM THE SCOTTISH BORDERS

Hard Lives for Poor Reward

Colin Whittemore

First published 2017

Copyright © Colin Whittemore 2017

Published by
Old Pond Publishing,
An imprint of 5M Publishing Ltd,
Benchmark House,
8 Smithy Wood Drive,
Sheffield, S35 1QN, UK
Tel: +44 (0) 0114 246 4799
www.oldpond.com

A catalogue record for this book is available from the British Library

ISBN 9781910456743

Book layout by Servis Filmsetting Ltd, Stockport, Cheshire
Printed by Replika Press Pvt Ltd, India
Illustrations by Colin Whittemore

THIS book is about the never-ending uphill struggle against the adversities of nature and the seemingly unrelenting state of depression in the Scottish Borders countryside. It is the real story of the people behind today's farming ways.

The narrative comes from historical evidence and the personal memories of Borders farmers.

The author acknowledges with thanks the time given in conversation with those who are fondly recorded in this book, and with many others. Other information is sourced from his own work and research on the social history of the farms and farmers of Howgate and Newlands and recorded in formal detail elsewhere.

About the author: Colin T. Whittemore, FRSE, is Emeritus Professor of Agriculture, University of Edinburgh.

Contact: colin.whittemore@btinternet.com

To Edinburgh

Penicuik

Auchendinny Mains

Newbiggin

Pomathorn

Pentland hills

Springfield

Howgate

Newhall

Ravelsyke
Venture Fair

Auchencorth Moss

Leadburn

Whim

Kingside

LaMancha

Linton

Grange

Sunnyside

Wester Deans

Maobie

Whitmuir

Noblehouse

Halmyre

Grassfield

Romanno

Docoot

Eddleston

Whiteside

Fingland

Moorfoot
hills

Noblehall

Newlands

To
Biggar,
Lanark
and
Carlisle

Callands

Flemington Mills

Scotston Knowes

Drochil

River Lyne

Castlecraig

To Peebles

Contents

1

The Vale

*T*HE *stories of Scotland's farming families begin with the making of Scotland's farms. These were times of innovation, of the sweeping aside of the old ways. The farming revolution brought the excitement of progress and the despair of dispossession.*

How had it all come to be? In the old landscape there were few farms. So we need to go back to that time; before the Clever People came. As for a date – a watershed – then let's say 1745, as good a date as any to talk of rebellions and revolutions. So what was going on before then?

Newlands vale folds into the form of a gentle curving glen from Howgate to Drochil with Romanno at its nub. It is neither large nor particularly spectacular – grandeur can be safely left to the Northern Highlands. Through the bottom

1

flows the River Lyne, which snakes purposefully, 500 feet above sea level, from valley side to valley side. The hill tops, rounded and swathed in short hill grasses, rise to twice that height, with a few heathery knolls peaking at over 1,200 feet.

In the early 1700s, Newlands was a populous place, busy with families tilling the land for their staple cereal grains, feeding animals for their meat and clothing, and digging peat for fuel. Their cottages were rubble-stone built, with roofs of turf, stone or thatch. The unit of life was the family, the work was the land, and the social structure a cooperating community helping itself with all those tasks that take many pairs of hands; tilling, harvesting, herding, killing.

The people lived *in* their Newlands landscape, not *upon* it. As often as not, a part of their sustenance would come directly from their natural environment, not just from the agrarian one. In those times the 'wild' landscape yielded a rich bounty of seasonal food for the ordinary people who were scattered through the countryside: fish, meat, eggs, fruit.

The salmon came up the Lyne from the Tweed every autumn to spawn in the pebbled pools of the upper reaches. There the smolts vied with the brown trout, grayling, sticklebacks, freshwater prawns, oysters and eels. The riverbanks were rich with wild-fowl.

The wetlands and bog ponds beside the river nurtured duck, snipe, goose, quail, gull, moorhen, coot and dabchick; all good for human food. Care was needed, however. The dark deceitful swampy flats of sphagnum moss grow cotton grass, asphodel and rush on the sides of the Dead burn and the Black burn. The first flows south to the Lyne while the second flows north to the Esk. Both are well named, flowing languid by hidden pools in the peaty wastes through which they wander. These places will swallow up any beast or human that does not watch where they tread as they make their way across the moor.

Higher up the slopes of the vale's sides there were trees scattered openly in the upland pastures. Here and there, not yet robbed for fuel and building materials, were denser woodlands hiding plenty of wildlife and plump-breasted pigeons living among the birch, pine, alder, beech and oak trees. Small herds of deer ventured out into the moor from the woodland fringes. The woodland itself, come autumn-time, also offered generous bounty; fungi, hazelnuts, blackberries, raspberries, blueberries. These were important supplements to cereal grains for family survival.

The hill tops and upland slopes either side of the glens above the woodland line were covered in tough grasses; fescues, bents, matgrass, interspersed with molinia clumps. The open grazings were abundant with flowers and herbs; harebell, eyebright, bedstraw, trefoil, vetch, pansy. In this, undercover, hid partridge, grouse and black cock. Above, eager for the chicks, circled hawks, buzzards and a pair of eagles.

The big houses all had dovecots, as well as rights to the deer. These dovecots provided sanctuary through the spring and summer to semi-tame doves that would nest and rear squabs. Through the winter, the 'doocot' – as it is known in Scots – would provide fresh meat for the kitchen.

Rabbits – once farmed in warrens, now escaped – were to be found wild in profusion and easily dropped with a well-directed stone hurled from a sling-shot. A fine meal they made. They could be dug out of their burrows, of course, with help from the dogs; but such practices were not sustainable – the doe needs peace to produce the next generation of dinners. Rabbits, for some families, were a prime and essential source of meat. There were a few hares on the high ground, but the hordes that would come later to plague the Valley had yet to arrive.

* * * * *

Looking down the valleys, the dominant aspect was one of openness; cultivatable space with houses and byres. The drier parts of the lower grounds, those naturally draining into the burns that fed the river, were cultivated by the people. The soil was good; clay on the north-facing slopes, and looking south the land is lighter; kames of glacial sands and gravel.

In 1745, the tipping date for change in the Newlands landscape, there were just short of 1,000 people living in the glens from Drochil to Romanno and Romanno to Howgate. The major settlements (touns) each had a score or so dwellings close by a larger heavy stone-walled bastle house, built by the laird. The touns had within them their own mix of relatively few extended families carrying well-known names, and also the many who are nameless, known only by their given name and perhaps, as a mark of respect, the nature of their employment.

Between the touns were the but-an'-bens – two roomed cottages – some with a byre attached for a cow, a pig or a few hens. These were spread apparently randomly over the landscape; they were built of rubble-stone and earth with roofs of turf or thatch. In them lived the cottars. The cottars paid a trivial rent for a length of rig – cultivatable ground – which was tilled to feed the family. The rigs would tend to lie parallel, and many cottars would have more than one rig, running side-by-side. It is these same cottars who will come to pay the ultimate penalty for the making of the farms.

By the 1700s, the threats of raids from the Border Reivers were mostly gone. The threat, if any, came from the religious divide; the people of the Valley followed in the tradition of the Protestant Dissenters, while from the north the Catholics were ever-plotting with the Francophile Jacobites to return the nation to Rome and the true faith. It was 1745 when Bonnie Prince Charlie marched through the vale going south. One year after that, the good 'Sweet William' Cumberland would be marching north to put a finish to all that pretentious Popish nonsense.

The valley had little history, for little was needed. Besides, there was nothing to write about. What was past was the same as what was to come. The present was for always. Generations came and went, each family to their own trade. Nothing happened in the valley – save only life and death.

History would begin only when a few of the enlightened gentry of Edinburgh decided that a fashionable thing to have might be an estate in the country, and a fashionable thing to do might be to improve upon what nature had for millennia thought best to provide.

Before the Enlightenment – often thought to have begun with the 1707 Act of Union – the families in the valley had simply got on with it. What preoccupied the people in the communities of the Lyne valley was working the land. The valley was full of families each surviving (or not) off what they grew on their strips of cultivated ground – the rigs.

The staple food upon which these families depended was bread (barley, wheat and rye), porridge (oats) and bannocks (flat oat bread). The porridge was eaten hot or cold. Porridge could be kept overnight and wrapped in a cloth for the next day's lunch. Primarily these people were arable (cereal) specialists, not livestock keepers.

When not actively attending to their plots, the cottars employed themselves in supporting trades; miller, smith, carter, stonemason, woodsman, carpenter, huntsman, weaver, peatcutter, tailor, bootmaker, tanner, potter. It was the *land* that those living in the Lyne glen depended upon for their lives. The rigs ensured a meal every day for the cottars and their families. Life was, however, no idyll. Most people were hungry much of the time. The living was hard, a bad winter meant that more of the children would die.

Casting a careful eye over the Lyne valley today, where presently the lower ground supports only sheep grazing in badly drained rush-infested pasture and the upper slopes are plastered in conifer tree plantations, one could be forgiven for finding it hard to believe that most of the lower ground, and much of the inclines, were, in the 1700s, arable cropped – tilled, sown and reaped. Oats, barley, wheat, rye, kale, potatoes, flax. The cottages had gardens for green and root vegetables.

There were chickens and pigs around the houses and in the touns. Sheep were kept for textile wool, and the old ewes slaughtered for winter meat. Milk was stolen from cows suckling calves to make butter and cheese. The pigs provided pork to cure, but most importantly pigs could turn cereal grain starch into wonderfully rich fat. Lard was essential to provide a medium for cooking as well as a vital source of energy to fuel a day's manual labour. Scotland did not grow many sunflowers or olive trees!

Barley made weak fermented ale. Meat was preserved with precious salt from the pans tended by the coastal monks where the tides came in over the rocky foreshore and the water captured to evaporate in the rays of the sun. Much of the higher ground was shared commonly as sheep and cattle grazing. Nobody thought that the first use of land was to make money. Its purpose was to provide directly sustenance for those living within it.

Things were different in Edinburgh. The Union of Crowns and then of Parliaments helped to make rich a new class of Scottish nobility. Money came from the benefits of English trading; from the colonies in the form of sugar, tobacco and slaves, but not least from the Industrial Revolution in Scotland itself; textiles (home-grown wool, imported cotton), manufacturing, coal, heavy engineering.

What to do with all this wealth? Perhaps a hobby, a pastime. Something cultured, a benefaction to society. Something enlightened, knowledgeable, clever. Something that a shrewd lawyer or merchant could be proud of, but at the same time would turn into a penny or two. How about improving the lot of those poor wretches living in the hinterlands of Edinburghshire? A rural estate perhaps? Howgate, Romanno, Grange? All well within a half-day's ride from the city.

There was serious money to be spent.

At first it was sport that brought the moneyed classes to the Lyne Valley in the age of enlightenment – wildfowl on the bogland, game fish in the rivers. They bought up the big fortified houses and developed them into pretty country retreats.

But then a grand and clever idea came up from England. The Members of Parliament learnt of it on their trips down to Westminster. It was called 'land enclosure'. A reformed landscape. Along with that came, through the late 1700s, the construction of purpose-built farm steadings, controlled rotation cropping, turnips and the supplementation of the soil with fertiliser. Along with that came food production; the feeding of the industrial working classes in the towns. Along with that came raised rents. And along with that came profits. The land wasn't just working to feed the people; it was working just like every other of the new industries – for the financial gain of the proprietor.

These were the new merchanting, legislat-ing and military families that sprang from the Enlightenment. Those such as Clerk at Newbiggin,

Islay at Blairbog, Montgomery at Coldcoat, Cochrane at Grange – the people with the grand ideas and the money to see them through. They wrestled the landscape from nature and remade it to their own design. They made farms – business units each with their own employees. The cottar communities that had inconveniently occupied the same lands were broken apart and summarily redistributed. An enforced diaspora that would populate the British colonies overseas with hard-working god-fearing Borders folk.

2

The Reverend Findlater

*T*HE *happenings that brought farms and steadings – the enclosures – were recorded faithfully by the minister at Newland's kirk. He was a man of parts, well-renowned. He did not volunteer for the job, he was commanded.*

At the very time when the first farms were being constructed in the Scottish Border country and the resident inhabitants were being cleared off, there came upon Sir John Sinclair of Ulbster (Member of Parliament for Caithness, economist, and agricultural scientist) the notion of a Grand Scottish Statistical Account. This would record everything that was going on in Scotland; parish by parish.

Responsibility to write and submit the Statistical Accounts was offloaded onto the parish ministers. The first account took place in the late 1780s, and the second in the mid-1800s. These were seminal to the future admin-

istration of the Scottish countryside and the alleviation of destitution.

The ambitions of Sir John Sinclair were common with those of the Church, both seeing the Statistical Account as the Lord's work. The Reverend Robert Wallace, moderator of the Church of Scotland, even in times of the terrible insurgency of Charles the Catholic Pretender could think only of how better to provide for the needs of the ordinary Scottish country people whose stomachs were empty and whose children were dying from pestilence. Sinclair's grand design had the same focus; to determine the extent of the poverty throughout Scotland and to put in place actions to address the plight of the poor rural communities.

The parish minister at Newlands and Grange was Charles Findlater. He was particularly well qualified to do the job. He would have known Sinclair; both having a deep interest in the 'science' of farming. Findlater was born and bred in the neighbouring parish of West Linton, he wrote authoritatively on farming matters. The fine Manse at Newlands lies above the Old Kirk, on the lane that winds up the hill to the hamlet and tower-house at Whiteside. Findlater was there from 1777 to 1835; years that would also define the enactment of the Agricultural Revolution; the building of steadings and fields, the application of enlightened science to farming, and the clearing of the common people – the cottars – from their run-rigs.

Charles Findlater's father, Thomas, had been settled through the patronage of the Duke of Hamilton into the living at West Linton. The congregation was not content at their minister being foisted upon them against their will, and in 1731 chased Thomas ignominiously out of the village, setting up a Secessionist kirk of their own. After a period of riotous upheaval, during which the troops were called, the troublemakers (mostly female) were marched to jail in Edinburgh, 20 miles distant. Thomas appears to have been placed into a neighbouring parish for a while before reinstatement, but his congregation now covered only a part of the parish.

Charles himself was put into Newlands parish by William Douglas, who as the Earl of March owned the lands west of the Lyne. Charles' appointment was also opposed, but it seems he was ultimately accepted. Secessionist Burgher kirk congregations were now at both West Linton and Howgate. Charles, however, pursued a learned, liberal and enlightened ministry.

The terrors of 1745 were still fresh in local memory. The fervently Protestant parishes of

Howgate, Newlands and West Linton were only a day's march from the Palace of Holyrood in Edinburgh, from whence Charles Stuart and his Roman Catholic Highlanders had set forth to march south to victory and the restoration of popery in England. The courageous sheriff-depute James Montgomery of Macbie (who was a formidable land reformer and steading builder, and owned a good deal of Findlater's parish) had bravely called the militia to arms to defend estate, faith and family. He was, however, shrewd enough to make his rallying cry only after the danger was well south past Carlisle. Charles Edward's retreat to Inverness a year later, by the grace of the Lord, left Newlands unscathed. The local parish considered Culloden to be a 'blessed victory' and celebrated it with an inscription carved into the stone lintel of a new barn nearby at Grange farm thus;

William Duke of Cumberland. Liberator and Defender. Culloden Moor. 16th April 1746.

Charles Findlater himself was respected by his parishioners for his agricultural advice as much as for his pastoral duties and his keeping of law and order. In those times the minister was absolutely at the centre of all parish activities; guidance on matters spiritual being as important as guidance on matters agricultural. Feeding the stomach and feeding the soul were equal concerns.

Findlater expounded on the best ways of milking dairy cows and the growing of turnips. He was deeply impressed by the benefits of land drainage using specially dug trenches and ditches; witnessing first-hand the transformation of the fields at Boreland across the valley from his manse. The landowners and tenant farmers of Newlands were his friends and associates as well as his parishioners.

In addition to his inputs into overseeing his Parish's farming revolution he was, like all good ministers, expected to baptise the children before they died, educate the ones that lived, divine the sins of the sick, give succour to the dying, the bereaved, the hungry, the poor, the disadvantaged and the dim-witted, and to render the lunatic into appropriate asylum. He (by law) presided over a court of discipline that admonished the drunken, exhorted the dancers and cavorters to mend their lewd ways, and remonstrated with those who did unnecessary tasks on the Sabbath, such as cow-skinning. The men who fornicated were admonished in closed session and asked to consider improving their behaviour. Erring women who were bold enough to allow these (tempted) men to fornicate with them were humiliated and disgraced in front of the whole Sunday congregation. The rights of the Kirk's Holy Communion were for the time being withdrawn from such women, which caused them great distress lest they died before their ways were mended.

The farming geography of Findlater's parish had already been well depicted by the most accomplished military maps made after 1746 by Major-General William Roy of Lanark. These maps show that more than one-third of the parish – that being in the valley bottoms and lower slopes – was successfully under the plough and in cultivation, while a further third was fine upland grazing for sheep and cattle.

The valley was densely inhabited by many cottar families conscientiously working to feed themselves by the cropping of fertile and giving land, which was divided into strips (or rigs) running side-by-side with furrows between to allow water to run off. This said, times and climate were hard and many of the cottar families were destitute – hence the main reason for Sinclair wishing a statistical account. The last third of the general's map showed bog and moss, while on the hill tops there was common ground and rough moorland. Of trees there were few.

Through to the 1500s it appears that the population of the countryside was in reasonable balance with the landscape they lived in, but during the next two centuries the climate turned cold, wet and sunless, bringing crop failures and hunger among the people. These bad times were made worse by the clearing of the woodlands to make way for more cropping land and to provide fuel, charcoal and building materials for Edinburgh's town tenements and the shipyards at Leith. The countryside, rather than a place of bounty, became treeless, harsh and barren.

Changed times, however, were what Charles Findlater reported to John Sinclair of Ulbster. Findlater rejoiced in the revolutionary activities of the new lairds; the Lord Islay of Blairbog (Whim), Baron Clerk of Newbiggin in Pennycook and Lord Chief Baron Montgomery of Coldcoat (Macbie). These men all pledged themselves to the improvement of their estate and the repair of the depredations of their predecessors.

In 1715, Dr Surgeon Alexander Pennecuik of Newhall Romana, who wrote widely of rural matters, had described with disdain a social structure where the landowners in possession of more than half of Scottish Borders land had little or no proprietorial concern for their estate and small interest in farming – their wealth being obtained from activities in the city such as banking, law or commerce, or else in the army or navy.

In those former years, the Scots tenant was held captive to one-year run-rig leases. There was no incentive to improve the land that, cherished by neither landlord nor tenant, fell to impoverishment. Because the poor cottars could pay little in rent, the landlords were disinclined to spend money on land and buildings. Thus the land would lie unnourished, and the cottars' damp rough-stone hovels fall to ruin.

The valley that Findlater found when he took residence in the Newlands manse was a poor and menial spot. The place he left when he retired was prosperous. Much of this was due to the farming knowledge and advice given by Findlater himself, who not only understood well the need for drainage and trees, fertiliser, lime and crop rotation, but was unremitting in his exhortation of all in Peeblesshire to take up these innovations.

Findlater, however, generously ascribed the heaven-sent change in the valley's fortunes to the Lords Islay and Montgomery. He would say – wise diplomat that he must have been – that it was by *their* application, imagination, innovation and investment that where cottars' hovels had once stood and where diggers of peat had once lived, now were to be found industrious farms; farms sufficient in size to employ hired hands.

Findlater waxed eloquent about the blessings that were the farmsteads, enclosed fields, drained soils, fine trees and the means to spread upon the crops dung and lime. He saw as a good thing the removing of the cottars and the peat-diggers who could barely feed themselves – never mind others – and their replacement with diligent tenants selected from those classes of person superior to the common people and thereby having the wit to farm the new ways and thus be able to pay good rentals from what was becoming a most bounteous landscape. Findlater was a strong advocate of another new idea, the 'savings bank', but overall he took the view that only where there is profit to be made of it, will an enterprise flourish and be cared for, both by landlord and tenant.

Findlater was so taken by his local lord that he dedicated his submission to Sinclair's Statistical Account to none other than Lord Chief Baron Montgomery.

* * * * *

One of Montgomery's chosen tenants for his new farm steading being made at Noblehouse was an 'inn-keeper' and portioner from West Linton, a Mr George Dalziel. Dalziel was the first person to grow turnips in his rotation, thus enabling stock to be 'over-wintered' rather than slaughtered in autumn. He folded his sheep on the turnips, containing the beasts with nets. This was a disaster, as the breed in question was the Linton Blackface, which had fine sets of horns on both rams and ewes. To prevent entanglement he had to saw off the horns, although sheep-proof walling round the fields was to prove a more sustainable option.

As a portioner, Dalziel would have traded in livestock, wool and leather, and owned a number of strips of land (rigs) around the parish. He would have taken menial rents from the cottars. With the clearing of the people off the land and into the cities he would have been looking for recompense. To become the secure tenant of a single large integrated piece of land (a farm indeed) with a fine house and steading would have appealed. In the closing years of the 1700s he would have been able to watch Noblehouse being built from scratch by Montgomery.

For the new tenants in the new farms, the revolution of industry in Lanark and Peebles and even Edinburgh was an opportunity not to be missed; the city would need the country to produce food for the workers in the mills. Food for which they would pay. But it was not immediately apparent to Dalziel that his choice to become a tenant of the Macbie estates was wise. Watching a house being built in foul weather is fraught enough, but a whole farm is a thousand times worse! Noblehouse under construction was, by all accounts, a sea of mire.

Montgomery had deemed that the house and steading be on the south side, sited above the wet of the valley bottom. There is a burn running down, with ample opportunity for an upper pond and lade to drive a wheel. But that burn runs to the Dead burn, which is well-named for it makes its way idly to the Lyne River, by way of bog and moss. The decision to have a burn running close by the steading as a source of water power was a good one, and common to nearly all the farms made in the valley.

Building a new steading required a veritable army of men. Local labourers – many none other than the erstwhile cottars themselves now put to more useful work – but also strangers from the west, even beyond the town of Lanark. The incomers must have been threatening to the locals. Being hungry, they were willing to work hard. Their accents would have been strange and their way of speaking difficult to understand. And then there would have been the fear of what might happen if, after the work was done, they did not go away. Being themselves dispossessed from their own homes in the west, they might not be inclined to return there, but to stay in the Newlands valley. As, in the event, some did.

The field enclosures were made according to the lie of the land and natural boundaries, but tended to be of ten to 20 acres each. Between the enclosures, trees were planted in strips some score of paces wide; birch, oak, beech, Scots pine and the Duke of Argyll's favourite, the larch from across the northern seas. Each side of the plantings, deep ditches were cut with smaller channels dug laterally across the open fields to join the ditches. Into these, stones were slung, then covered again with brush and sandy soil carted from the north side.

Dalziel found it hard to believe that rubble-stone walls fastened with lime mortar could make a house for himself and a steading with byres and granaries for his farm. But fortunately there were examples to hand where steadings had been built earlier and were now settled down. One such was that built by the Lord Islay of Blairbog at Deans as a dairy farm for his tenant, Thomas Stevenson.

The house at Deans was rather fine, with slated roofing, dressed stone at the frontage and a fine stair. The kitchen was separated from the living quarters and the retiring rooms were quite apart. Something much to be appreciated by Mr Stevenson's wife and family! At the top of the house, within the roof, there was a small stair to spaces where two housemaids were kept. Next to the house was built a dairy place for making butter and cheeses. The steading yard was broad enough to throw out the dung, and there were three rows of stone buildings all the same; two rows of byres for milking cows, horses and stock, and one for the families of the herd, the dairy-man and the ploughman. The granary was of two stories, with a horse gang for powering an oat mill.

Stevenson's dairy cows were productive; cheeses and butter going into Edinburgh for sale at good profit. Drains were built at Wester Deans from the outset, ensuring the best grass in Edinburghshire.

Noblehouse was to become, and for that matter to remain until this day, one of the most productive farms in Peeblesshire. Within were two tracts of south-facing gently inclined lands that the dozen or so cottar families who farmed there had called Plewland and Sunnyside – a bit of a giveaway! Their but-an'-bens had been cleared;

the rubble being tidied into heaps to make all possible space available for tillage. Some stones were used to make drainage channels, while others were added to those from the quarry for the steading walls and field boundaries.

Dalziel, and others of the new generation of tenants, took to modern farming ways with a will appropriate to the culture of enlightenment that prevailed in society at the time. The landscape had, in no more than half a century, turned itself from having the purpose of feeding those that lived upon it to feeding those that did not. Farms from now on would be there for people with no connection with rural life; the miners and the shipwrights, the industry workers in the foundries, the weavers and spinners in the mills. This was a very large new idea; one that could only have been realised if the land was made to produce twofold, nay fourfold, more.

At the root of the revolution were the smart notions picked up by the likes of Montgomery from his English associates in the Parliament. These can be simply stated as follows.

First, divest the local small-time farmers of their leases so that their strips and patches of land can be amalgamated into fields of good and practical size. Next, enclose these fields with a wall, hedge or fence to separate crop from livestock. Then drain, lime and fertilise. Last, vary the usage of the fields year upon year in a four-course rotation between grains, winter roots, grains under-sown with grass seeds, and finally the grass cropped or grazed with livestock for three or four years before being ploughed for corn again.

Taken all together the result was a fourfold gain in the yield of grains and a doubling of the output from the livestock. More than enough to pay for the increased rent that would be charged. At the same time, new machines were invented that allowed a larger scale of operations to be completed faster and with fewer people. With less of the population employed on the land, more could be employed in the factories. A virtuous circle for both farming food-producer and industrial employer!

Not all of those who were cleared from the land at Noblehouse gave themselves up to self-imposed slavery in the factories of the Industrial Revolution. Nor did they wish to become mere hinds – lowly serfs at the beck and call of new owners of the likes of Dalziel, who now presumed to have rights over the very land from which they had been summarily evicted. Some, however, did stay on, giving up a hovel that was falling down for the delights of a cottage row and (if fortunate) a big vegetable patch at the new steading.

Those of more independent spirit took the option of going to the colonies, where a new – rural – life was promised. Many did not make it, but those that did, ended up helping to create the farming landscape of North America.

Families from Scotland's Borders, as like as not, would find themselves packed up and walking away to Glasgow dock, with the blessing of the local laird. Those from Newlands would not have been the only ones from Peeblesshire to board boats such as the *Jane Duffie*, a three-masted sailing barque, bound for Quebec. Many would end up growing grain in Canada's open prairies. Place names from Scotland are liberally replicated from Nova Scotia to Vancouver. Lord Selkirk had done his best to help the unfortunates by setting up the Red River Settlement, near Winnipeg. It was good grain country, but quite the coldest, dullest, flattest place to be imagined, literally in the middle of nowhere; a far cry from Newlands and the soft uplands of Scotland's sweet Border country.

3

John Carstairs Esq

*T*HE *flight was scheduled for 18.30 on the 15 May 2015. We were to meet with the pilot – and all the paraphernalia that is so much part of ballooning – at Penicuik, on the green where 200 years earlier Mr James Jackson's market garden had been. The weather was fine and sunny, but the breeze was wrong, taking us as it would toward the city. So we returned to the place from which I had started out; we would inflate the envelope and take off from Broomlee at West Linton.*

The basket lifted off, gaining height quickly over the blasts of hot air from the gas burners. Then all was quiet and we were drifting. First below was New Kaimes House, built by Major Edward Thomson of Callands as a wedding present for his younger and rather less introverted brother Captain Ronald, proprietor of Halmyre Mains. Then over the cropped fields of Noblehouse, toward the mess of little paddocks and dwellings that had once been the fine country house of Macbiehill, before it was knocked to rubble and made into roads. We passed directly over Islay's drainage scheme which provided the necessary beginnings for what was to be an extended country garden at Whim house; the gardens that never were. But the drainage ditches are still to be seen, even though the area is covered with brush, trees and regenerating bog mosses. Blaircochrane's rush-infested grasslands hove into view and beyond ... to the north ... what in the name of Christendom is that?

What it was, was some 400 acres of green grass, cut clear from the browns of the huge expanse of Auchencorth moss that completely surrounded the plot. The place had all the appearance of having been mystically conjured into exist-ence. The boundaries were excised with the precision of a boning-knife. One side moss, the other side pasture. Within that Shangri-La lay a handsome white house of Georgian-style build, with some outbuildings that must have been the last remains of an old steading. Across the patch were lines; drainage lines. We could see it from our vantage of the wicker basket; laterals into mains, mains into ditches, ditches into the Black burn, bound for the Esk River.

What was that all about? It was, well, not believable – incredible. We passed on toward Kingside and Mount Lothian, but I could not take my eyes off what I could scarce believe. It was in the middle of the moss; no road ventured past it. There was one lane in – a dead-end to nowhere.

Later that evening, champagne all done, I looked out the OS map. There it was. I had never noticed it before – never had cause to. There was no footpath or way through it or past it. It was surrounded completely by the moss of Auchencorth. It was a farm – optimistically called Springfield.

I was curious. Who made it? Why there? What was it for? When?

I turned to my tried and tested companion, The Transactions of the Royal Highland and Agricultural Society; *the volume did not disappoint. The sluice gates opened and out flooded the story. I was entranced.*

* * * * *

Edinburgh's Highland Society was formed in 1784 in response to the need to improve farming practices in the Highlands where many were at that time starving. The Transactions of the Highland Society *began in 1799 and are a rich mine in which to quarry the progress of innovation and improvement through the years of the Agricultural Revolution. By 1835 the* Transactions *were of the Highland* and *Agricultural Society. My first hit was in the* Transactions *of 1826. I had searched for the word 'drainage'. A certain John Carstairs is awarded the Highland Society of Scotland Honorary Gold Medal for his work on the topic. Then in 1843–45, Carstairs gives an account of one of his many fertiliser experiments*

– '*Account of experiments with guano, bones, and some special manures on mossland*'.

So here is his story in as much as I am able to determine it.

That part of the Auchencorth moss that lies at 800ft above sea level north of Rosemay and south of Halls on the Leadburn side of the Black burn is particular in the bleakness of its bog. A family of cottars eked out an existence living in a low windowless turf-roofed hovel. Their main source of sustenance was from the plentiful supply of deep peat that they dug from the 20ft thickness of flow moss, before setting it to dry for their own hearths and for those of others living in the Howgate community. In 1750 or thereabouts, Sir John Clerk, in waxing lyrical about his Penicuik properties, manages (not without conspicuous effort) to find at least one thing good to say about the moss. Being uncultivated and marshy, its improvement 'gives work for my tenants and servants'. He asserts that the swamps are 'suited for sport; wild geese, partridges, quail'.

To the mossy wastes of Auchencorth came one of the foremost characters in the history of the reclamation of agricultural lands. He was a pioneer of land drainage. He possessed those qualities obligatory for such work; spirit, stubbornness, inventiveness, sufficient funds and a desire to be recognised by others as being worthy of note.

Surveying his newly acquired domain, he observed a cotton-grass and sphagnum bog of acid pools and raised peat, fit for neither man nor animal and worth 'not a sixpence per acre in rent'.

One wonders why he came at all. His name was John Carstairs Esquire. He died in 1854 at Springfield Farm aged 66 of cirrhosis of the liver. Springfield came into John Carstairs' possession in the early 1800s. At first he may have intended simply to obtain revenues from digging peat from the flow moss. But instead, he decided that he would follow the example of the pioneering Campbell, Duke of Argyll, and Montgomery of Stanhope. He set-to with a passion to create a farm out of the wilderness. So progressive was he in his farming practices and his scientific examinations of the new ways of husbandry that he became a national and international expert on matters of land reclamation, drainage and improvement. His work is quoted as *the* shining example in the agricultural writings and texts of the time; he is applauded for his work, which becomes a major element in the section on land drainage in the first *Chambers' Encyclopaedia – a Dictionary of Universal Knowledge for the People* (1860).

First a roadway was led in from Milkhall. Then a lattice of usable paths was forced across the bog. The horses required plates upon their feet to prevent them sinking into the mire. Belts of woodland six to 12 trees wide were planted in straight lines, following the example of John Clerk. These formed square enclosures of 20 to 40 acres. This extraordinary pattern is quite evident on contemporary maps (see, for example OS 1854, surveyed 1852). Within each main enclosure, fields were formed by the setting up of earthen dykes and hedges. Then the fields were levelled and drained. As these operations were mostly done with a hand spade, Carstairs must have been a wonderful employer of local labour! What was not drained by him remains flow moss to this day. Surveys from the 1840s show an area of around 500 acres at Springfield. Two-thirds of this are laid into symmetrical fields with dykes and tree-breaks between, clearly in arable cultivation, while the remaining third is enclosed and roaded ready for the next phases of reclamation.

Carstairs' drainage technique was to force through a network of open ditches leading down into the local burns (one of which is the Black burn). Into the ditches ran drains dug to some four feet in depth and spaced nine to 18 feet apart. The trench was back-filled with gravel and stones (out of which he had sieved the sand, which was itself spread on top of the moss). On top of the gravel went brush and rushes.

The water ran out of the mire through the stone conduits and the moss dried.

The land was now ready for the next phase, which was to improve the physical and fertility characteristics of the soil. This involved the adding and ploughing-in of copious quantities of sand, lime (from Macbiehill) and clay. All of this was carted in. For this purpose, Carstairs made his wagons to run on transportable wooden rails; being pulled by those horses with plates on their feet.

Sometime in the midst of all this, Carstairs must have built a steading for his arable operations and a decent house for his wife, Marion Brown of Kirknewton, who he married on 12 August 1834 at age 46. There would be no issue. By 1840, Carstairs had built a prodigious steading; by far the largest in the whole of the Howgate area. But the rather fine house built slightly apart from the steading is not yet evident.

By the 1840s, John Carstairs had one of the

most renowned and productive crop farms in all of Midlothian and the Borders. His yields in terms of tons per acre and quality of product were phenomenal. Where previously had been bog was now prime arable land growing clover, oats, wheat, barley, hay leys, potatoes, seed potatoes, turnips. He even had apple trees! An oasis indeed!

It wasn't just the pioneering land reclamation work for which he became known, it was also for the subsequent improvement of soil fertility. He led the way in using the developing science of agricultural chemistry. He implemented the new ideas of rationalised scientific method – the doing away with traditional practice and folklore and the application instead of logical thought and analysis.

His experiments were impeccable in their design and interpretation. One such is reported in the 1843 RHAS Transactions – he got a gold medal for it! He had a control treatment to establish his baseline; to this he added nothing, just planted potatoes. He found, unsurprisingly, the yield to be low. On an equal plot he added 'bone manure', noting that yield increased by one-and-a-half times. To his third plot he added ammonia salts – with a similar response. His fourth plot was bone plus ammonia plus guano (concentrated bird droppings, dug from South American sea cliffs made by generations of bird colonies and shipped into Leith docks). This increased yield to twice that of the control. His fifth treatment was a heavier dose of guano. This doubled the yield again (four times the control). The sixth plot he fertilised with a liberal dose of farm yard dung – same result as the guano. With such a willingness to take careful steps in observing responses to trial treatments, it is no surprise that as an innovator applying the new ideas of the Agricultural Revolution he gained a renowned reputation!

The puzzle however, remains. Why did he do it? Was it folly? Would it all endure?

Carstairs reckoned Springfield worth 'not a sixpence' in rent per acre when he came in. On that basis, about half of £1 (10 shillings) per acre would have been paid for its purchase. The estimated costs of the reclamations are given as £15 per acre. By the middle of the century, 40 or so years later, the reclaimed lands on the farm were worth a rental equivalent of £2 per acre; an 80-fold increase. He would just about have got his money back. Unfortunately, however, he did not live to enjoy the profits of his labours. When Carstairs came into Springfield, it supported one family. When he left it, it supported ten families (not counting himself as a gentleman of sub-stance) and on top of that was the casual labour coming out of the local hamlets.

Just before his death in 1854, John Carstairs, as was customary, drew up his will. The whole of Springfield farm is be left to 'my beloved Spouse, Marion', for Marion to manage 'according to her pleasure', but only providing she remains unmar-ried and conducts herself satisfactorily. Should Marion marry again then her rights 'shall cease … in the same way as if she were naturally dead'.

In the event, Marion did *not* conduct herself 'satisfactorily'. She died of 'old age' at 97 in 1912 at Springfield. She is noted as 'Widow of 1. John Carstairs Landed Proprietor, 2. Robert Murray Farmer'. Robert Murray, by-the-by was her nearest neighbour! Marion died a woman of prop-erty. In 1906, she is listed as owning a number of houses and lands in and around Howgate in addition to Springfield – which in 1909 she let to one Thomas Lambie of Castlehill, Manor Valley, Peeblesshire.

* * * * *

Without John Carstairs at the helm, the fortunes of Springfield dwindled over the years. The cycles

37

of agricultural depressions that kicked in around 1870 dragged the place down. Following the Second World War, the ploughmen, stocksmen, dairymaids and labourers were discharged one by one. The arable fell back to grass. The farm has since grazed cattle and sheep looked after by one man with a stick and a sheep dog. Rush incursions show that the land is now again getting sodden wet; tumbling back to the way it once was. And all around, the mire of Auchencorth moss is still there; brooding, threatening, waiting.

4

Venture Fair

*E*IGHT *miles south of the town of Edinburgh, on the way out of Howgate village, going 'south' toward Linton there was a settlement – still is – dating from the 1600s. In the middle of the 1800s, when we get some sort of record of its recent inhabitants, it comprised three sets of rubble-stone buildings. The first was the 'proper' farm steading of Mosshouses. The second cluster was what had been the cottar's dwellings at Venture Fair and were now cottages for tied labourers – the farm hands. The cottars had originally come as independent workers tilling their own meagre plots, but now worked at Mosshouses and Walltower in lieu of rent. Lastly there was a byre and a mean two-roomed but-an'-ben stone shack set back between the farm and the cottages, called Nethermosshouses. The few acres of land here were good, within sight the substantial farms of the Clerk estates at Halls and beyond, and up the hill the old-established holdings of Kingside and Herbertshaws.*

Next after Mosshouses began the boglands of Auchencorth moss; wicked and dangerous. But out of it, recently cut by John Carstairs, was the heroic 300 acres of Springfield. Springfield was the most productive farm in Edinburghshire until the depression of the mid/late 1800s hit the farming community. Perhaps it was hard times that brought William Elliot north from Selkirk toward the lure of work in the town of Edinburgh. He stopped his travelling when he got as far as Venture Fair.

The Elliots represented a very particular sector of the Scottish farming hierarchy. They aspired to be above the common labourer – the hired hand dependent upon the whims of their masters and, it was said, little better than slaves. The hired hands were fired at will, and tended to move employers – and the living quarters that went with them – on an almost annual basis. Their wives having to make some sort of home out of what the previous resident had just been pleased enough to leave behind them.

The likes of the Elliots held a middle ground; above that of the common labourer but unlikely themselves ever to become owners of proper family farms in their own right; although it was exactly that to which they aspired.

The same pattern of life is still to be found in

the Scottish farming scene of today in the shape of the 'Scots shepherd'. The herds expect the owners of their flocks to give them a free hand in the shepherding, and indeed to share the owner's land for grazing their own flocks. Many of these 'self-employed' farm workers take on contracts with the bigger farmers and stock owners as it pleases them; harvesting, shearing, gathering and dosing, tupping, lambing. Those same workers will value their freedom and independence to be difficult, to choose to refuse. They will have patches of rented ground and grazings on short-term lets here and there.

Back home at their cottages there will be a vegetable plot; perhaps in previous years a pig or two and chickens. They are to be seen weekly at the common stock markets, buying and selling; but rarely borrowing. They are self-reliant with the expectation of making their own beds, and when things get tough, defiantly lying in them. When things are good and the money is flowing, they do *not* expect to be troubled by the taxman.

It would appear that William Elliot was an 'innkeeper'. But in the early 1800s this was a loosely used term that included many who made and sold beer. As beer was a (microbiologically) safe everyday drink for everyday people, brewing was commonplace. It would seem, however, that William had had something at Selkirk to sell, for he came with money in his hand to buy. The Elliot family possessed Venture Fair in their own right.

Venture Fair was not a farm, but neither was it a worker's menial hovel. It was a substantial one-storey cottage row (two dwellings) with a byre attached, and another substantial outbuilding for livestock, or possibly a working horse. With the buildings came an acre or two – enough to graze a cow and a handful of sheep, and to grow the all-essential vegetables and winter kale for the family. It was enough to be self-sufficient.

The previous incumbents, as was usual practice, left the premises that they vacated in bad order. As flitting was usually a result of hard times, assets would be stripped. The idea that income would be reinvested in repairs and maintenance remained an unfulfilled dream. William would have been met with a lot of reinstatement work to be done. Farming was depressed. He and his family would have needed to work for others as well as for them-selves. But there were work opportunities around at the neighbouring farms, or if all else failed, in the local rural industries; quarrying, mining.

William brought with him his elder son Robert and his younger son Alexander. With Alexander came Alex's fecund and dutiful wife Janet. Janet bore children in regular succession, five of them living to adulthood; William, Janet, Elizabeth, Jane and little Alex. It was Robert who seemed to do the farm-worker bit. Alexander had to go and get work elsewhere; first at Cowan's Paper Mills in Penicuik, and then as a platelayer on the new railway being pushed through from Hardengreen to Coalyburn (the coal mines at Macbie). Alexander lived to the age of 64, dying in 1889.

In 1901, the family, like every other family whose livelihood depended upon the land, hit hard times. To raise cash, the Elliots sold half their house to the well-to-do Pows in the big house at Walltower, a few hundred yards down the road. The families squeezed into the other half. But not for long. Little Alex and big sister Janet – known as Jessie – took up the let of Nethermosshouses next door. The cottage had two rooms and, better, the byre there would hold a couple of milking cows. There were nine acres of grazing at Nethermosshouses – enough for the cows, a few sheep, hens of course, and the vegetable garden. It's Jessie who does the work about the place – she has ambitions to build the farm up. Alex has a job on the railway; following his father.

In the crowded rented conditions of the

tenements at Templevale in Howgate village lived
Betsy Walker. Betsy was pleased enough to get
out at the age of 25 and join Alex (aged 34) in
wedded bliss at Nethermosshouses. They had
four children. Their life was just about settling
into a family routine when the landlord decided to
sell the place. Sitting tenants are not much of an
inducement to potential purchasers, so Alex, Betsy
and the four kids were ignominiously thrown out.

They go back to Venture Fair, which must have
got quite cosy! Alex's other elder sister, Elizabeth
– Lizzie – is still there, looking after dairy cows
and taking in lodgers. These were mostly exports
from Edinburgh's 'poor houses', young hope-
fuls looking for work at farms, mines, or in the
big houses. Lizzie seemed to have four or five of
these unfortunates shoehorned in at any one time!
Somewhere along the line the Elliot family got
back that part of the property they had offloaded
to the Pows, and by 1911 they were back in full
possession of all of Venture Fair. By then the Pow
family was in disarray, the second of its members
being committed to the lunatic asylum in 1907.

Betsy had ambitions to be a farmer, even
though they had left Nethermosshouses and
joined the family regrouped at Venture Fair.
There was only McGill's acre in-bye at Venture

44

Fair, but there was a byre that would take six milking cows. Alex rented fields down at neighbouring Milkhall and on the Kirk Glebe. At first light, and every evening, Betsy went to get the cows and walk them up the road to her byre for milking. Through the winter months they stayed chained up in their stalls.

From her cows, Betsy had milk to sell. The village was well supplied from the established herds at Walltower, Howgate and Howgatemouth, so Betsy shipped her milk into Edinburgh. Her ten-year old son, Sandy (who was born in 1909), was in charge of the horse and cart, so it was he who delivered the milk to Leadburn station to be put on the seven o'clock morning milk run back from Peebles to Edinburgh. Betsy needed to feed her five living children, but her enterprise was too small to support the family. Too small even with the rented acres, the six milking cows, the two breeding sows and dozen piglets growing in the pig sties, the duck flock, and the 50 hens she had producing eggs to sell down in Penicuik. Alex had to keep his full-time job as a platelayer on the railway, as well as tending a garden vegetable plot ever-increasing in scale and productivity (thanks to the manure from all the animals).

There were no 'utilities' at Venture Fair. Not

even a water wheel. Not even water! All the water for the family and the livestock (dairy cows drink huge amounts!) had to be hand-carted in buckets 150 metres from a spring in a field across the road. Over a few weeks in 1930 Alex used his platelayer's shovels to dig down in his garden till he was deep enough to strike water. When he did, he put in a well, set a hand pump on top of it and catapulted Venture Fair into the modern age! (Electricity would not be getting there for another 30 years; and the best winter heating available was from the warm bodies of the milking cows).

In the years before the Second World War, the Elliot children (there were eight before Betsy was finished; five girls and three boys) were just let to run wild about the country – but that is what country children were meant to do. In due time, each of the family were sent their separate ways so as to make space in the overcrowded cottage.

Betsy and Alex's children were put out to work as soon as was permitted. Until they became teen-agers they worked at home and at local farms, but could not be formally 'employed'. This did not stop a huge proportion of seasonal work on the bigger farms all being undertaken wholly by, or with the help of, 'underage' children – potato and fruit picking, haymaking, harvesting, lambing, shearing and so on. The good thing was that the family was able to keep the kids out of the coal mines and lime workings.

Annie and Sandy got good jobs at Springfield farm; she as dairymaid, he as a horseman. They both lived in the farm bothy; which was a lot worse than mucking in with the family at Venture Fair. Sandy got out as soon as he could to a better position as a labourer at Walltower farm, right opposite Venture Fair. Sandy worked at Walltower for the rest of his life, being awarded the Highland and Agricultural Society's Gold Medal for long service in 1981 (aged 72). Soon after starting work at Walltower, Sandy had been struck by a stone on his ankle. He was too busy working to bother with the infection that ensued. The resulting septicaemia nearly killed him – to save his life they cut off the bottom of the limb. He spent a lifetime of farm labouring with one-and-a-half legs! While working at Walltower, Sandy went to live at Ravelsyke, one of Walltower's cottages.

Jessie did well securing a job in domestic service in a big house in Edinburgh town. Jim went to work as a horseman at Halls farm. Young Lizzie followed him to a job there as a dairymaid. Lizzie got engaged to Peter Abernethy of Howgatemouth farm in the village, a lengthy engagement that lasted 12 years! Finally, they married and Peter got a job as signalman on the railway. A good number of local people were employed first building, then maintaining and running the local railway operations. This was surprising as it was but a single track, serving mostly to ship coal from Macbie, and a few passengers from Peebles!

Lizzie came across the march fence between Halls and Howgatemouth and set-to to make a go of farming what was a perfectly respectably sized mixed family farm. Lizzie had waited, but Lizzie had come good; she had made the jump from farm-worker to farmer.

Unfortunately, the farm's deeds were not in Peter's name, but that of his absent elder brother. The elder brother came home and decided he wanted the house and farm and shoved Lizzie and Peter out. They went to jobs at Kirkettle, by Roslin.

Grace went off as a kitchen maid for a little while with the Clerks at Penicuik House (who, their big house having burned down, now occupied the substantially re-configured stable and carriage block). She left there to take up with one of the horsemen at Walltower, Jimmy Glasgow. With the coming of the second farming revolution after the World War II, and the tractors that went with it, Jimmy found that his services were no longer required. Jimmy and Grace went off to live in the roadsman's cottage at Silverburn, from which Jimmy looked after the eight-mile stretch of road between Carlops and Flotterstone.

The last-born of Betsy's family at Venture Fair was Jeremiah in 1929. Upon reaching the age of 14, like so many other country boys in the district, Jeremiah walked everyday up the steep hill to Kingside farm to serve his apprenticeship as a ploughman. This he accomplished ten years later. On the strength of his completed training he married Evelyn, a town girl from nearby Penicuik, and took her off to Marchwell to manage the pig farm there. For him and Evelyn it was serfs' work, so he bettered himself with a farm manager's job in Fife for a while before coming back to Marfield to look after the huge beef feedlot enterprise there. In 1970 something better came up. Farms at that time needed to reduce expenditures, and there was a massive shift of labour away from the land. Jerry was one of the ones to go. In the event he ended up in a nice job with the water board, and a rather fine house for himself, Evelyn and his tea-pot collection.

Peggy's ambitions were necessarily limited

– she stayed home with mother Betsy. At the time of Jeremiah's birth back in 1929, Alex was still the head of the family with a strong hand in what went on at Venture Fair. The 1930s were a time of deep depression for farming; there was no money in anything. Only those who owned their farms could survive. The Elliots did not; their fields were rented and payments had to go out even when no income was coming in. Old Alex decided that they would have to stop spending. The consequences were, as was the case with countless little farms up and down the country, that the cows were got rid of and no money was spent on maintaining the fabric of the buildings. Farm and house at Venture Fair fell into dereliction. The remaining land that the Elliots did own, McGill's acre, was used to park caravans – second homes for families in Edinburgh and Leith with a desire to spend their holidays in strange faraway places, like Howgate. Venture Fair visibly shrank. The cottage row became shorter. The byre roofs fell in. The living spaces in the cottage row diminished. The imperative for the children to move out ever-strengthened. Things got progressively worse through the war years.

Lizzie and Peter had been dispatched to Kirkettle by Peter's elder brother in the early

1950s. Around then, the living conditions at Venture Fair had become intolerable. Betsy and her daughter Peggy finally left the family home, or what was left of it, and went to join the others at Kirkettle. Venture Fair was abandoned to its own devices – namely to fall to ruin.

Sandy decided he would vacate Ravelsyke and build for himself a new house (of prefabricated concrete) at Venture Fair on top of his father Alex's pride and joy – the vegetable garden. Alex had died in 1946. It was perhaps fortunate that he had not been a living witness to the tragedy. The caravans, now a significant eyesore and annoyance to neighbours, and no longer earning any money, were got rid of. The cottage row at Venture Fair tumbled ignominiously into its final days – as a shelter for Sandy's hens. His mother Betsy regrettably was a living witness to all of this. Given her long and eventful life at Venture Fair, it must have been deeply wounding for Betsy to see what had been not just her family home, but also her farm – her life's work – crumble into a pile of rubble lying where had been the two cottages full of the Elliot families, the byres with their milking cows, the horse in its stable, the pig pens, the sheep fold, the duck house, the poultry shed. A family's farm; now a stone heap.

5

A Few Hens for a Bit of Pin Money

*P*RIOR *to the 1950s almost every farm in the valley had a few hens. Sometimes these did more than just produce eggs for the family farm kitchen; eggs were sold into neighbouring villages and towns. The flocks producing eggs for sale could be quite large; even as many as 200 or 300 birds. It had been thus for 150 years, with one notable exception. In 1915 the biggest poultry farm in the whole of Britain was built at Romanno. It was a very strange business. The people in this story are shadowy, their behaviour and their motives mysterious, but what they did was quite extraordinary.*

Frank Edward Adams was a highly successful entrepreneur, innovator and engineer. He had businesses in London, Cheshire and elsewhere. Among other things, his factories bent steel and made metal boxes and cans. He was an enthusiastic fly fisherman and used to come up to Romanno for the trout fishing in the Lyne. The impending sale of parts of the Romanno estate came to his notice, so he bought a large part of the village, including the red-stone Temperance Hotel, the village houses, the cottage row and the tweed mill. With it went a few acres of surrounding ground and, of course, the fishing in the river flowing through the property.

There is no clue as to why Adams should wish to set up a poultry breeding business, nor why he should think his holiday fishing spot – located as it was between the estates of Coldlands and Coldcote – a particularly auspicious location to set down multiple varieties of breeding birds whose lineage originated in tropical climates! It may have been his engineering skills that alerted him to a business opportunity.

He was the very first person to grasp the concept of *factory farming*.

In the early 1900s, poultry scratched in yards, milk was drawn from cow's udders by a sucking calf or a dairymaid's gentle hand, pigs lived in sties with outside runs, and chicks were hatched from eggs nestled under a broody hen – just like every other bird in God's world. Mechanical implements had come to the farming of crops 100 years earlier, it is true; horses pulled ploughs and machines to sow seeds, and waterwheels turned gears and shafts that threshed the sheaves of corn. But none of this had touched the farming of livestock.

Adams had learned of a new-fangled bit of kit from America; an incubator to keep hens' eggs at exactly the right temperature to hatch them out as chicks. Instead of eight eggs under one hen, an incubator could hatch hundreds of eggs. And, what's more, hens only came broody once a year in the spring, whereas the mechanical incubator warmed up at any time of year when the heaters were fired up. This was the first step in a recipe for animal exploitation that was both endless in scale and continuous in time – an egg factory. And F.E. Adams saw it. The old linen and tweed mill was stuffed full to bursting with Adams' mechanical incubators.

But Adams did not go in for egg production, not at first anyway. He, being the astute business man that he was, realised that there was more money in selling stock for others to farm, than in farming himself. What Adams did was to breed egg-laying birds to sell to the many thousands of small poultry production units that resided at most every farm up and down the country. Adams was aiming to supply all of them with his stock. He sold day-old chicks, he sold young pullets, he sold point-of-lay hens. They were shipped by road or by rail out of Broomlee station. Once other farmers began to pick up on the automated incubation of eggs, he began to sell fertile eggs from his 'improved breeds'; the company went international!

Frank E. Adams put his brother, George H., in charge of the poultry operation, setting him up in the old temperance hotel – a fitting place, given that George was a most pious man. Why George should have been side-lined up into a tiny Scottish hamlet rather than helping to run the prosperous engineering company (which, given the date, must have been going full speed in support of the war effort), is not at all obvious. Unless, of course, Frank Adams had come to the conclusion that the success of the company would be best served by his brother's removal far away from it!

Neither of the Adams brothers knew a thing about poultry, so they appointed a young man exempted from army service on the strength of a maimed hand. His name was Thomas Linkie. Who, it might be gathered, did know something.

Adams and Linkie set about improving the quality of laying birds by the well tested methods of the sheep and cattle breeders. They decided what they wanted (which in this case was more eggs), then selected those female birds that in the previous cycle of lay had produced above the average, and mated them with cocks hatched out of eggs from only the very best egg-layer mothers. It wasn't long before the word got around that while your bog-standard farm-yard chuck would lay 80 odd eggs a year, an Adams and Linkie hen would lay at least twice that and more. The business was a runaway triumph.

Bearing in mind that an average poultry flock at this time was about 50 birds, a serious egg-laying enterprise would have 200 birds, and big units up to 500 birds, the statistics out of little Romanno in Newlands valley along the Lyne beggar belief.

Ten-thousand chicks were hatched from the incubators weekly (the running capacity of the incubators in the tweed mill was for 60,000 eggs). The eggs came from, conservatively, 7,000 birds in colonies scattered in breeding groups over 300 acres of ground. Adams and Linkie sold at least a dozen different strains of laying birds, but mainly Famous Romanno Black Leghorns and Famous Romanno White Leghorns. They sold turkeys and geese. Then they diversified into selling poultry feed (enriched with meat meal from spent cart-horses) to go with their birds. They built houses of titanic scale to accommodate 15,000 birds; these were used for rearing their point-of-lay stock. When they found that they could not always fill their orders at exactly the right time, they realised that they needed a cushion of overproduction of chicks. The males were shipped into Edinburgh and Glasgow to provide luxury meat for the upper classes. The females were shunted into specially built laying houses to produce eggs directly for human consumption.

At peak, Adams and Linkie had at Romanno around 7,000 breeding hens and 3,000 commercial layers (and the poultry feed business). This was by far the largest flock and biggest poultry enterprise in the whole of the UK. It would have staggered the ordinary farmer who with a mere 5 per cent of that number of birds would have thought himself to be 'a producer of importance'.

Ten years after its heyday, the whole lot was gone.

There are few signs remaining except the imprint of the brick founds of the six layer houses. The colony and rearer houses were wood and likely ended up as fire fuel. The dispatch department and feed mill had a new road-straightening scheme built over them. George Adams' house became a pub (he would have been appalled). The incubators fell to ruin and rats. A part of the mill tried to be a steak house, and then ultimately the whole lot was divided into flats. By 1924, Thomas Linkie had gone to Newark in England, and George had disappeared. He left behind in Romanno his two girls. It is difficult to be sure of the extent of their disadvantage at being thus abandoned. They lived quietly together in one of the red-stone cottages until their demise. With them for a while was a man who became totally mute, but whose origins are unclear.

Through the course of the 1920s, farming fortunes in general had begun to take a turn for the worse, not really picking up again for another 20 years. Trading would have slowed up. Egg prices slumped in the face of cheap European imports. On top of that, the mere number of birds all on a single site would have been an invitation to disaster. The breeding colonies were in huts on open ground where there would have been a build-up of parasites, intestinal worms and diseases such as salmonella, coccidiosis, E. coli, tuberculosis. The place would most certainly have become 'fowl sick'. Disaster was bound to follow.

The final nail in the coffin of 'Britain's biggest poultry farm' was the fate of Frank Edward Adams himself – the industrialist backer to the whole enterprise. The *London Gazette*, which lists company dissolutions and bankruptcies, records his companies going down one after the other from 1926 through to 1929 when his last remaining business crashed and was wound up.

It is a matter of record that, today, the Newlands valley hosts – at Whim, Easter Deans and Blyth Bank, another huge poultry farm – Glenrath Eggs, the business of John Campbell. Glenrath have well over a million hens in the valley, each one laying 300 eggs every year. Adams' big layer sheds held 500 birds each; Campbell's hold 32,000. John Campbell, like many highly successful industrialists is keen that you understand he is other than that; merely a simple shepherd from Argyll no less.

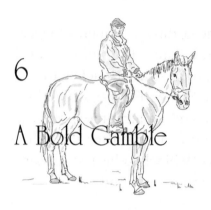

6

A Bold Gamble

THE Pow family of Walltower, Howgate, who had come there to farm in 1850 had tired of it by 1880, and were keen to get the whole place let out. The formal let would be for Walltower farm together with the little steading opposite at Ravelsyke. The let is recorded as from 1892 to 1912 to Messrs William and George Cairns, whose business office was at 189 Fountainbridge, Edinburgh – close by the sprawling McEwan's brewery complex there. The Pows would stay on at the big farmhouse. The farm grieve would have the cottage that was part of the Walltower farm buildings, while if the new tenant wanted somewhere to live, there was a cottage set just above the steading at Ravelsyke.

William and George Cairns' father, John, had a number of town dairies located around

Fountainbridge. These were large byres housing milking cows. Once incarcerated, the cows never saw a field, being fed on concentrates and by-products. A major feed was brewer's draff; the barley waste from the brewing industry. This had to be carted from brewery to dairy. Unsurprisingly, the business that was passed on to William and George was lucrative and expanding; trading in animal feeds as well as milk, while also operating as general haulage contractors. In 1897 The National Fat Stock Club awarded the Cairns brothers a 'Special prize for best dairy cow fed on condiment'.

The Cairns were rough, ready and getting rich. There was, however, no element of their work that could be thought of as in any way genteel. They would have been seen by Edinburgh society as somewhat coarse. Carting draff was a dirty, smelly occupation, while conditions for the dairy cows in the town byres were brutish – nothing that gentlefolk would want in any way to be associated with.

Pneumonia and tuberculosis were rife among cattle. The *Transactions of the Highland and Agricultural Society of Scotland* record that William was called as a witness on 18 May 1887 to recount his experiences of lung disease to an expert committee looking into cattle pneumonias. Tuberculosis is a highly infectious and lethal disease that, through the consumption of milk contaminated with the organism, carried off large numbers of Edinburgh's children in the nineteenth century.

Later, around the turn of the century, the Cairns would be involved in the first UK importations of a specialist breed of Dutch dairy cow – the black-and-white Friesian, now the most populous milk cow in the UK. Part of the reason for this was to protect against breed wipe-out in Holland due to the ravages of foot and mouth disease.

With the increase in their wealth, the Cairns brothers developed ambitions – ambitions above their station. They saw their way to acceptance in Edinburgh society as being through joining the Establishment at the posh end – the horse-racing set. But the Cairns were not gamblers, they were businessmen. They would not be betting on the horses, they would be breeding and selling them. They would become racehorse owners and breeders. Animals were, after all, things they were familiar with.

The arrangements at Walltower were that the farming side would be handled by the grieve, while George would look after Ravelsyke and the breeding of the horses. They employed John Kerr (from the stables at Crichton, Pathhead) as head groom. He proved a most excellent choice. While George was in the engine-room, brother William was front-of-house, doing the rounds of the race tracks; dressed for the part, in smart town coat, kidskin gloves, bowler hat and cravat.

When they came to Walltower, the cottage row at Ravelsyke was largely abandoned. Farming was in depression. Farming systems were simplified and the labour force paid off. The cottages that had housed the farm workers would be turned into stables – Kerr's Yard – with accommodation for John Kerr at the end of the row. George would use the small house set above the steading for his own family's use. He was keen for them to enjoy life in the fresh, safer, country air.

George brought his wife Margaret and their two sons to Ravelsyke in 1892. Six years later Margaret died of hepatitis. Margaret's sister Janet helped raise the two boys, and in 1900 bore for George his third son (named after him). They shared their time between the house in Leamington Terrace and the one at Ravelsyke, but it was evident that the dirty work was at Fountain Bridge, and the fun times were at the Ravelsyke stables.

Elder brother William dressed smart and went to the races; glad-handing potential buyers for their horses, passing pleasantries with the county gentry. William set about climbing Edinburgh's social ladder; spending happy days among those whose number he so wanted to join.

George, in contrast, took to the country life and looked after the brood mares in the stable block. William did the buying of the foundation stock brood mares. Clearly the haulage and town dairy business must have been making money, because William bought well. At Ravelsyke there were eight or nine mares, each expected to earn their keep by being put to the stallion as two-year-olds, and bearing yearly thereafter foals destined for sale. The horse business was to be fun, as well as being the route for advancing Cairns' social ambitions. But above all it would need to make money. It was not in the Cairns family nature not to make money. Besides, if the best thoroughbred mares were to be bought into Kerr's Yard, then they could only be afforded if the best prices were gained from the offspring. If things went well, the best blood females could be kept home, and the Yard would gain its own reputation for excellence in its racing quality.

The brood mares at Kerr's Yard included Lady Cashier, Kalydor, Morocco, Counterpart, Palmary, Lady Strathairlie, Ballochmyle and Julia

Gaylord; the last three being home-bred. More than a dozen stallions were used to cover the mares, but certain favourites were used more than others to establish a thoroughbred type that would characterize the Cairns' bloodlines.

William's favourites were Queen's Birthday, Galashiels, FitzSimon and Tarpoly. But his star stallion was Diplomat. He covered, at one time or another, every one of the Ravelsyke mares.

The horse-breeding partnership of the Cairns brothers was formally set up in 1890. They started winning races 'under rules' immediately and continued through until 1906. For the most part they ran youngsters – two-year-olds – in 'selling races'. The purpose being to do well, get an offer from another owner looking for a young prospect, and sell forthwith. William Cairns' racing colours were blue, with red sleeves and white cap. William and George Cairns were hugely successful at breeding racing horses. They could show the gentry how it should be done. A point not missed, and greatly resented, by their more long-established 'betters'.

William Cairns had 51 winners – £6,000 worth, a huge amount in those times – in the course of his time as an owner. However, as an owner-breeder, the prize money was a means to an end. That end was selling his protégés as best he may.

All the horses running in the colours of 'William Cairns' were listed as bred by 'Messrs Cairns'. He was racing home-bred – Ravelsyke-bred – horses. George had had a winner under his own name back in 1898, but after that it was all "Owner – W. Cairns; Breeder – Messrs Cairns".

The top stallion bred by George Cairns at Kerr's Yard, Howgate, was Cumnock Lad, who won four races under the colours of William Cairns.

Cumnock Lad was sired by Diplomat and was bred out of George's favourite mare, Ballochmyle. Ballochmyle herself had been foaled with John Kerr there beside her at the Ravelsyke yard in 1899. William got her mated as a two-year-old to the best sire he knew; Diplomat. She was too young and thankfully no foal came to term. In 1902, Diplomat covered her again. Ballochmyle foaled Cumnock Lad in 1903.

In 1904, William Cairns sent Cumnock Lad for training at John McCall's yard, Westbarns, Dunbar. The colt was raced the full five furlong distance on the flat no less than ten times as a two-year-old, the first being at Ripon in Yorkshire. After that it was Hamilton Park, York, Eglinton, Stockton, Ayr, Edinburgh, Newcastle, Liverpool and Manchester. He was placed third at Ayr and Edinburgh and won at Eglinton (in July of that year) and then three in a row; winning at Newcastle, Liverpool and Manchester, all in the space of five weeks. In the first of these, John McCall rode him himself, before passing the reins to E. Wheatley.

The time was ripe; William Cairns put Cumnock Lad up for sale by auction. Nothing that either John McCall or John Kerr could say would make him change his mind. Cumnock Lad was to be sold while the going was good. He was bought by Major Joicey, a Northumberland coal owner, for a purported 1,700 guineas. Joicey ran him 18 times all around the circuits of England over the next four years, but with three exceptions only, Cumnock Lad never placed again. Joicey sold him off into Belgium, a country renowned for its love of a good horsemeat steak. Nothing more is heard of Cumnock Lad.

In 1905, W. Cairns had 12 winners. His young horses, 11 of them, were entered into 65 races, mostly on tracks in the north of England and in Scotland. All these horses were bred by Messrs Cairns at Ravelsyke. Cairns mostly used John McCall of Dunbar as his trainer, but in the year of victories, William Binnie (also Scots born) at Malton, Yorkshire, was also involved. The year

1905 was hectic. One in five of all the Cairns' races produced winners, and all were home-bred: a formidable record. When it came to breeding racehorses, the Cairns brothers could show the Establishment a thing or two.

Ballochmyle had two more foals after Cumnock Lad. They were also colts by Diplomat. One, Cumnock Hero, was gelded and sold out to Germany. The last colt born to Ballochmyle was in 1905. He was a poor specimen, either born dead, or dying soon after. Ballochmyle was sold suddenly late in 1905, to be shipped unceremoniously to an unknown destination in South America. This was the first sign that all was not well at the Yard.

After 1906, silence. Nothing, no record of breeding or foaling. No race winners. No sales of promising youngsters. All the brood mares and all the colts and fillies were disposed of by the end of 1906. The enterprise at Kerr's Yard was dispersed – cashed in. There is nothing to be heard on the racing circuit of 'William Cairns, racehorse owner', nor of 'Messrs Cairns, thoroughbred horse breeders'.

It transpired that William and George Cairns had put their affairs into the hands of one Robert Millar. Robert Cockburn Millar CA was a partner in the well-renowned firm of Barstow and Millar, Accountants, at Edinburgh's North Bridge. Millar was a highly esteemed member of Edinburgh society, being made JP. He was a founding member, and president of the Edinburgh Society of Accountants. He specialised in bankruptcies and liquidations.

For some reason or another, 1906 found the Cairns brothers staring ruin in the eye. Their companies were stone broke. With the help of Robert Millar, the brothers avoided the ignominy of appearing in the *Edinburgh Gazette* under the heading of 'Liquidations, Bankruptcies and Sequestrations'. But in so doing, they *did* have to turn all their liquid assets into cash. And among their most valuable saleable assets was the thoroughbred racing stock mares and foals at Kerr's Yard. It broke their hearts.

There was no doubt who had won in the end … the Establishment.

George Cairns died in 1912 of a heart attack while loading a cart with brewery waste at Fountainbridge. Things must have come to a pretty pass for the erstwhile breeder of winning racing horses, leaseholder of Walltower Farm, proprietor of Edinburgh haulage and dairying businesses, to be *himself* howking animal feed

69

from a pile in the yard and up onto the back of a cart.

* * * * *

What of George's youngest son, named after him? He too would take on the Establishment. Like his father, it would kill him. The Establishment would break young George just as it broke his father. But unlike his father, young George would win.

George Cairns Junior died on 23 March 1962, in Trafalgar Hospital, Oakville, Ontario. He was proud to have been a Canadian, but prouder still to be a Scot; committed, bristly, red-headed. He insisted on his voice being heard. It did him no good, but the world's community of racing horses have much to thank him for.

He had spent the last ten years of his professional life making trouble for those who exploited racehorses, mostly through the medium of doping them in one way or another. It was not the corruption (as he saw it) among the Establishment's racing fraternity that got George's goat the most, it was the way the horses were treated.

Young George was born in 1900 at Leamington Terrace, Edinburgh, but spent his early life at Ravelsyke, part of Walltower Farm, Howgate, where his father and uncle stabled their brood mares. George's childhood was among racehorses and their grooms. He had his own pony from the age of four – a Shetland-cross that would also pull a small trap. George Cairns was a horseman through and through.

When he came back from the war he could not make a go of things. The tenancy at Walltower did not appeal; he decided that a new start was needed. At the Port of Glasgow, on 19 April 1922, he embarked upon the Canadian Pacific steamship *Metagama*. He arrived, alone and penniless, in immigration reception, Quebec. He took up farming, at St Mary's, Ontario, but did not stick to it for long. He applied to Toronto Veterinary College, which had just moved to nearby Guelph, gaining a place there by examination. He spent the rest of his life as a horse vet.

Dr George Cairns DVM, MRCVS came to have an international reputation. Not only did he have his own practices, but he also became a respected and senior member of the veterinary faculty at Guelph. Being a part of the university 'establishment', he was considered appropriate to be appointed to the Ontario Racing Commission as their senior veterinary adviser. This was a post ritually seen as a sinecure for somebody ready to bend to the will of his employers – the horse-owners and trainers. They had got the wrong man! George was a country-bred Scot who would be beholden to no man! He saw his job as looking to the welfare of the horses, not the Establishment. It had been his experiences at Ravelsyke that gave him the courage of his convictions that would dominate his life in Canada.

As a senior academic and practitioner at the University Veterinary School, George taught equine diseases. The title of his principal course was 'Sporadic diseases, hygiene and surgery of the horse'. George became internationally respected as a horse surgeon. He travelled widely to operate on valuable racehorses. He was called in to treat cowboy film star Tom Mix's horse, Tony 'the wonder horse'. During the Second World War he served with the rank of captain. In 1941, George Cairns is recorded as overseeing the shipping of mules from Georgia through Nova Scotia to join the transatlantic convoys running the U-boat gauntlet. George went with them. The mules supported the Second Polish Corps in the Italian Campaign at the battles for Monte Cassino.

What was to separate George Cairns from the others, however (apart from being a feisty Scot), was that he had a novel approach to what being a horse vet should mean: not just curing sick horses, but also looking to the welfare of those not so sick. He had seen plenty of animal abuse among the racing fraternity in Edinburgh. His mentor at Ravelsyke, John Kerr, had shown him there was another way – kindness. George Cairns was well ahead of his time in considering animal welfare as a matter of legitimate concern for a veterinarian. It was this that would get him into *his* spot of bother with the Establishment.

George was appointed senior veterinary advisor to the Ontario Racing Commission in 1951. He was to the Commission what Thomas à Becket had been to Henry II. The 'veterinary advisor' had, until George, been considered as the Commission's poodle. But George believed more in the dogged pursuit of what was right for the horses, not for the horse-owners. He was bone-headed in his insistence that the administration of powerful painkillers to injured animals before a race was not in the best interests of the horses. If they needed that, then they should not be racing at all.

At the nub of it were the owners. They thought that they alone had the right to decide upon the fate of their property – the horse. Whether a horse did or did not race was their affair. George demurred. George was of the view that the fitness of a horse to be galloped competitively was a matter of the animal's *health*, not its *ownership*. A horse should be found fit to race, or otherwise, by an *independent* person trained to make such decisions; a *veterinary surgeon*. And George was a vet. He won through, in the end. But his battle with the Establishment lasted all the rest of his professional life, and shortened it.

The Commission overruled Cairns on his judgements of racing fitness of entered horses on more than 70 separate occasions in his first year of office. George snapped, resigned his post and went public with the scandal. The people who mattered in the state of Ontario, and his erstwhile friends and colleagues at the Veterinary School were appalled at George's brazenness. Their difficulty, however, was that they could not wrong him. He knew what he was about. Not just the horses, he knew about horse-owners too. He had seen them in action through his youth with his Uncle William around the Edinburgh race tracks. He knew all the tricks and fixes. Cairns knew when he was right. He pointed the finger of cruelty and corruption at the horse-owners. Not only were they doping the horses, they were duping the betting public.

The Ontario Racing Commission bought in other veterinary advisers to counter George Cairns' judgements. Unfortunately for George, these were his colleagues. He became 'less than popular' and was not made welcome in his own professional circles. George pointed the finger at them too, stating that their behaviour was bringing the veterinary profession into disrepute.

George Cairns raised the matter with the public press first in 1952 and frequently thereafter. The press loved it. The Establishment did not. 'The scandal of lame horses' made headlines. In February 1955, none other than *Life* magazine published an incendiary letter from George regarding horse doping and the complicity of the Establishment in it. The Veterinary Association hauled George up before its Disciplinary Committee. By 1956, George was taking on everybody, including his own profession. He had no allies in powerful places. But he was determined to win – for the horses' sakes; for John Kerr's memory and for the horses he had so loved through his boyhood at Ravelsyke.

In 1958, the Ontario Racing Commission backed down. By 1960, functional rules were being put into place outlawing unethical practices. The uproar generated by George Cairns' 'going public' had finally prevailed. The horses benefitted hugely. Two years later, aged 61, George Cairns died, like his father, of a heart attack. As his wife said, 'He took to his grave the courage of his convictions'. Nobody thought fit to write his obituary.

7

A Gentle Man's Estate

THERE'S no denying the Thomsons were rich. The family owned the biggest and best shipping company in Scotland – the Ben Line. A branch of the family – two brothers, Edward and Ronald – had land and property in the Newlands Valley; Edward at Callands and Boreland, and Ronald at Halmyre. Edward had Boreland House knocked down, improved Callands, then built Kaimes House as a wedding present for his younger brother (actually, for his new wife). Ronald became famous in the county; as Sir Ronald, Lord Lieutenant of Peebleshire and a great deal more.

Edward was a quieter soul. Shy, private, unassuming. How popular he was with the rest of the family one can only guess, for he spent much of his life buying and collecting – houses, estates, farms, motor cars, whatever – and then giving them away. He picked up a war-traumatised youth (whose father happened to be one of the last whalers to operate out of the harbour of Leith) from a garage forecourt in 1919 after a 'frank exchange of words'. He took the lad back to Callands and employed him to look after his fleet of heritage cars. After a lifetime of service, the 'youth's' retirement present was the very substantial farm, house and steading at West Mains, Castlecraig. Castlecraig itself had been earlier given away as a school for physically handicapped children, and Netherurd House as the Scottish Centre for Girl Guiding.

For some 100 years, through until nearing the end of the twentieth century, the 1,000-plus acres of the Lyne Valley in the ownership of the Thomsons was well looked after, but not worked hard. The farms did not need to make money, the Ben Line did that well enough. This was probably fortunate, because the years between the end of the 1800s and the middle of the 1900s were beset with farming depressions, low productivity, poor incomes and huge imports of cheap food from overseas. When Major Edward died in 1977 (aged 87), Callands was sold off to pay death duties, while Bordlands and Halmyre fell to Sir Ronald's son William, who continued the family tradition of gentle stewardship of the land; sheep flocks, beef and a little careful arable cropping for winter feed.

In earlier years, just like every other farm in the valley, the Thomson estates had had dairy cows. These were largely got rid of in the times of dismal milk prices and high costs following World War II, but in the 1970s, 'Young Mr William' wanted to start up a 'model' dairy herd at Halmyre, housed in a brand-new, automated, state-of-the-art milking parlour unit. EEC milk quotas, introduced in the mid-1980s, ensured that Europe would no longer be flooded with milk lakes, but in doing so rendered non-viable most of those farms that were not already large. The Halmyre herd was one of those. The dairy closed leaving only one other in a valley where 50 years before there had been more than a score.

In due course, the steading at Halmyre was left empty and a new steading built near Kaimes. At the present time, Halmyre is falling apart awaiting a bold developer to build bijou cottages on the site for Edinburgh city-dwellers aspiring to a country living. The steading at Callands had been converted to a square of houses a while before – the modern generation presuming the ordered arrangements there to be the idea of an inspired builder!

Meanwhile at Bordlands the walled garden still has one of its walls, but no garden. The Bordlands farm buildings are now also houses, made of good thick stone. Better to remain upright and used, than bulldozed flat like the big house has been. The stables, improved and upgraded by Edward Thomson in 1919 for his heavy horses, are no longer discernible. Which is a pity, because they had a story.

From the *London Gazette*, 13 September 1918. Piece referring to Captain Edward Gordon Thomson, Royal Scots, Royal Garrison Artillery:

Capt. Edward Gordon Thomson, RGA. For conspicuous gallantry and devotion to duty. This officer was in charge of a mobile section of his battery, where he received orders to retire, being at the time under heavy shell and machine-gun fire. His coolness and gallantry enabled him successfully to withdraw the guns, and he brought them into action in a new position. Later on, though most of his NCOs had become casualties, by his personal efforts and disregard of danger, he maintained the constant service of his guns.

Major Edward Thomson, MC, Royal Artillery, had been well-served by his heavy draughthorses,

upon whom he depended to pull the big guns. A few survived the apocalypse through to the armistice. Their fate at the war's end was either to be shot where they stood, taken off to be slaughtered for their meat, or – if lucky – given to Belgian farmers as workhorses. The cost of bringing them home could never have been justified.

Major Edward, however, was a tender man, who had been through a lot with his horses; he owed them. A sort of settling of a debt. Anyway, Major Edward was not letting his beasts get shot or eaten: his horses were coming home with him. And he had the means to get his way. After all, he owned a very large shipping line. So he got his horses back to Bordlands. They lived out the rest of their lives in luxurious retirement, loved, honoured and cared for.

One gets the notion that Edward Thomson was not much of a shipping magnate. He was just a nice man, who as a youngster (he was 24 in 1914) had been terrified and traumatised by his wartime experiences. Recognising that he happened to be rich enough that he could afford it, he modestly tried for the rest of life to be nicer to the world than the world had earlier been to him. It wasn't just the people around him who benefitted from his kindnesses – the land did too. It did not need to earn its keep. He did not need it for his survival – something that could not be said for many others living in the valley during the same difficult times. They needed to work their farms hard if they were to earn enough to keep body together with soul.

When Edward Thomson purchased the Castlecraig estates following the death of James Mann in 1941, he also acquired Scotston, and with Scotston came Scotston's Knowes. The Knowes, also called Whinnybraes (the banks were covered in yellow gorse), were a slopey, wooded, rough patch of banked lands running down the south sides of Drochil hill to the River Tarth. There was a farm there, let for generations upon generation to the Johnston family. Originally Scotston had been enclosed to make a farm by Captain Aeneas Mackay in the late 1700s. The Johnstons must have taken the tenancy sometime after the mid-1800s, just about when the agricultural depression set in. In 1911, the 'Johnstons of the Knowes' were Grandfather James, son James, his wife and seven children.

The family there in 1941, facing dispossession upon the death of their landlord and the dispersal of his estate, were to all intents and purposes indistinguishable from the family there when Queen Victoria reigned. In the event, the following year found them in full possession of the farm, now called Scotston Bank. They would not have had the means to buy it at anything near the market rate. One senses here the gentle and generous hand of Edward Thomson.

The Johnstons lived (and died) in a time vacuum. A couple of milking cows, a pair of farm cart horses (Rosie and Jean), a cattle-court with some fattening beef cattle, pigs in the pig-sty fed on household scraps, chickens scratching in the yard, sheep on the hill. The fields grew hay and oats to feed the animals over winter. In the house was old James, who had the best bed – by the kitchen range whose fire never went out. One of the children, David, had drowned in Castlecraig pond. The other brothers – John and twins Sandy and James – looked to the stock, the horses and the fields. House was kept by their sister Maggie. The twins were identical, and remained so all their lives – much to the confusion of everybody around. Maggie had her own room. The three brothers had a room too and lived all their lives at Scotston Bank with their three beds in the same one room upstairs. They spoke their own language, which nobody else could understand – it was likely a local form of lowland Scots; now dead, along with the Johnstons.

The family lived frugally, most of the time 'off the land' – their own eggs, bacon, mutton and milk. A liberal supply of rabbit, hare and trout supplemented the diet. Otherwise, in addition to porridge, home-grown turnips, kale and potatoes were the standard fare. Many families in the country had to relearn self-sufficiency in the 1940s, but not the Johnstons – they had known no different.

This is not so surprising – such ways of life were common within living memory. The house pig would be killed and dressed at home, and what was not feasted on fresh served up with blood-soaked oats was cured for bacon. Farming families would find wild rabbit and sheep's brains regularly on the menu.

When the Johnstons finally gave up the unequal struggle (they could never re-skill), the old steading was left and a big new one built a bit up the hill. The farmhouse was gutted and modernised. The Aitkens, who were at Scotston Bank after the Johnstons, also came to take on the land that had been Callands. Callands itself is no longer an estate. It is just a big house; one with a fine demeanour and a huge annual maintenance bill.

The Thomsons had followed a long tradition of beneficent landowners with a history of pouring

their wealth – made elsewhere – into farming. A characteristic of this charitable activity is that nobody expected any money to be made of these 'agricultural investments'. As the Duke of Argyll put it, referring to his revolutionary agricultural improvements at Blairbog: 'No, not a business venture, more of a whim.'

8

On the Hirsel

WHEN Tom Borland took on the 2,500 acres of Flemington hill in 1934 he had little inkling of what was to come. Wemyss and March Estates on the other hand were quite delighted.

The Earldom of March covered the titles of Lyne, Peebles and Neidpath. The Douglas family who held the title is iconic to the Borders. The titles came together in 1826, toward the end of the Agricultural Revolution when new farm steadings were being built and land enclosed into fields. Wemyss and March held vast tracts of Scotland's lands, not least in the marches of the Scottish Borders. In particular they owned the 2,500 acres that comprised the three holdings of Flemington, Whiteside and Fingland. These places were hill farms with only a little in-bye. In terms of agricultural value, they were about as low as can be got this side of heather moor. Borders hills are special, however; soft in shape, rolling, welcoming, grassy. They are nonetheless Scottish hills; demanding of respect, challenging and unforgiving in winter.

Flemington was not put there as a farm steading. It was a group of mill buildings – Fleemingtoun Mills – making use of water power from Flemington burn as it tumbled down the steep hill slopes, through the wooded gorge and on to the River Lyne. With at least two functional mills and workers' cottages to go with them, Flemington was an industrious place.

When the mills closed in the late 1800s, the estates of Wemyss and March were not much interested in putting capital improvements into Flemington. It was hill land, and hills are best just left to their own devices. Agricultural Revolutions, steading-building, making fields and so on doesn't happen on hill farms. Indeed, farming-wise nothing much happens on hills except cold, wet, storm and death. Although on a fine day, the vistas can be quite spectacular.

By the mid-1930s the mill had been long forgotten and the mill buildings were now used as a farm steading and cottages for the herds – despite being in two straight lines along the mill lade.

The main farm, with a few buildings, shepherd's cottage and most of the better grazing was up at Whiteside, where there had once stood a fine Borders Tower-house, glowering down upon Newlands Kirk by the river below. Fingland, for its part, was a high shieling with its own but-an'-ben and a small strip of workable ground. By the 1930s it was in as remote a spot as any shepherd's wife could wish not to be. But 100 years earlier it was alongside the drove road from Linton to Peebles and would have seen plenty of human activity as drover, journeyman, pack-pony, carrier, carter and cadger wended their way up the hill to climb to the slap in the high pass.

Tom Borland had learned his farming in Dumfries, South Africa and New Zealand. He was all set to go back to New Zealand and start his life as a dairy farmer when he got a flurry of frantic telegrams telling him to stay in Scotland as New Zealand had fallen into the grips of an agricultural depression. Of the choice of agricultural depressions, Tom unwisely decided to become part of the Scottish one.

John Dickson, Flemington's out-going tenant, had had enough of the unforgiving life as a hill shepherd and turned in his tenancy. Wemyss and March Estates offered it to the Cockburn brothers, but they preferred to work side-by-side (as farming brothers always do) running the vastly better farm at Kingside, Howgate. The tenancy at Flemington was going begging. Tom Borland, regrettably, did not walk by on the other side.

Hill farms are owned by neither title-holder nor tenant. They are owned by the grouse, the plover, the curlew, the trout, the deer, the heather, the blaeberry, and the grassy swards of fescues, molinia and nardus. The dominant grass landscape of Borders hills is offset by the occasional stand of Scots pine, much loved by kestrels. After all of that, there may be space also for a few hardy sheep and fewer still hardy cattle; those that can find a bite to eat and brave the harshness of the elemental wind and rain.

Given a strong constitution, a love of independence and a strange disposition, a shepherd may sometime be allowed to share in the privileges and vicissitudes of hill life; providing it is understood they are there on sufferance, and only temporarily. The privileges include the delights of a good spring lambing, the sweaty companionship of the sheep-shearing, and above all the freedom to walk the hill and arrange the day at no one else's bidding.

True hill – and Flemington is such with White

Knowe, White Hope, Drum Maw, Green Knowe, Hag Law and Wether Law – divides itself by the lie of the land into hirsels (Old Norse – places of safekeeping for sheep). A hirsel is a sweep of country, usually about enough for 400 or 500 sheep and one shepherd with his family. In some parts of Scotland this can be a number of thousand acres. At Flemington there were three hirsels; Whiteside, Fingland and Flemington – stretches of almost 1,000 acres each.

The hirsel is possessed by its flock of sheep. In this case the Scottish Blackface sheep; the only breed that has any chance of survival living in the fresh air at an elevation of 2,000ft in the depths of a Scottish winter. The flock on the hirsel lives there generation upon generation; flock and hill an inseparable unity. The sheep are not livestock, they are landscape.

The flock is hefted onto the hirsel. They live there, own it, rarely straying unless blown by storm or chased by dog. A heft is the place where an extended sheep family will live. The lambs learn family territorial boundaries from their mothers and will always be there or thereabouts. The land remains open; fences and walls are found only by the steadings. The hefted sheep are enclosed not by walls but by the inheritance in their minds.

Mere human beings are expected to follow nature's code of conduct for hill occupation. Upon a new tenant taking up a hill farm, after a little bargaining, the flock (and usually also that flock's shepherd) will stay with the hirsel. Unfortunately for Tom Borland, John Dickson had found himself short on cash upon departing Flemington and taken the whole Fingland flock to Peebles market and flogged the lot. There could be no worse a start for Tom. It would take him and Bob Lamont, the new Fingland shepherd, to breed more than five generations of a fresh flock before they got themselves hefted. Shepherding an unhefted flock of bought-in gimmers on an unfenced hill was a nightmare. Bob was forever fetching and carrying back his sheep from all points of the compass.

Tom Borland responded to the challenges of the hill for 50 years. Maybe in all that time he never fully came to terms with the idea that a hill could not be farmed. Farming is about disturbing the balance of nature to make surplus for the benefit of profit. A hill knocked out of kilter will give back only tears.

Tom had inside him a dairy farmer from the lusher pastures of Scotland's south-west, where farming was about controlling the land, pressing it for more, shaping it, making it work. The hill in contrast has to be partnered. The shrewd hill farmer asks the land what it wants, what is its will, and then goes with it. The hill shepherd is in no hurry. Resolute, yes. Philosophical, even more so; having every ambition to find peace with the world and to learn to live better within it. But no ambition to change it. The shepherd knows fine who is master and who servant. It would take years of herculean effort, pain and toil for Tom to be reconciled with Flemington's hills and to learn why it was that Border hill country breeds poets.

Flemington's wide high acres are blessed with tough short grasses. The natural equilibrium is a scattering of pines, broad sweeps of rough grass, some heather and in the wet patches rush, sphagnum and cotton grass. This balance is kept good by the animals that live there; the sheep and the cattle.

Scottish Blackface sheep and Highland or Galloway cattle have always lived together on the Scottish hills; they need each other to make a stable grazing ecology. Unsustainable grazing patterns will be exploited by insurgents and the hill will fall victim to bracken, rushes and gorse. To give up on the hill and plaster it with blanket forest is a gesture of despair and defeat that does landscape no favours. Mercifully, Flemington remains a grassy place, spared thus far, albeit encroached upon on three sides by regimented timber stands of Sitka spruce.

Old Bob Lamont, the shepherd who Tom Borland had brought across with him from the west, came to Fingland in the 1930s and stayed until his death in 1964. His son (Young Bob) was due to inherit the care of the Fingland hirsel from him. This was not to be, for in 1961, on a bitter February night, when the sleet was lashing down, Young Bob went out from Flemington steading to check that the cattle were safely sheltering in the woodland. His body was found in the Lyne water, maybe having first been bowled down the swollen Fingland burn.

Farming depression in the 1800s meant for hill shepherds a tougher life, less food, more infant death and shorter lifespans. But after the Second World War, Clement Attlee and the 1947 Agriculture Act, expectations were higher and alternative forms of employment more ready to hand. When things started to get really tough on the hill in the 1950s, a cut in wages was no longer an option; leaving the land, however, was. It was a simple matter of employing fewer shepherds. Not that there was any great clamour from skilled stocksmen looking for hill shepherding posts at that time!

Mary Lamont had got weary of the short summers with the long walks down and back from Fingland to the bottom steading at Flemington, and the long winters snowed in with Bob for weeks on end up at Fingland, where she had to make do on green kale grown in the vegetable patch there and stores of staples carried up by horse and cart before the storms set in. Bob too was feeling the strain of digging ewes out of snow drifts, throwing aside the dead and rescuing heavily pregnant ewes from going down with lambing sickness. In the 1950s, Fingland and Whiteside were abandoned; the few buildings left to fall to ruin.

Low productivity, depressed prices for hill lambs and wool worth less than the cost of shearing it off, meant that the Borlands had to economise. There were three ways out; intensify, simplify and cut staff. Putting more sheep on the hill and increasing the value of the lamb by going organic proved a double disaster. The hill did not like the increase in stocking and the market was not interested in higher-priced organic lamb. Simplification meant getting rid of the cattle, thus upsetting the delicate ecology of the grazing land.

Getting rid of the shepherds worked; reducing the number of families working at Flemington from five to one certainly saved a fist of money.

Tom Borland died in 1986, since which time Flemington has been looked after – struggled with – by his daughter Joan, the one remaining hill shepherd, the Border collies and the quad-bike. There are 1,000 or more sheep up there. The hill grasslands are giving way to invading rushes, bracken and gorse. There is no living left in it – at 2,500 acres, it just isn't big enough.

It is said, with some truth, that the Borders hill lands are unchanging, and to try to change them will end in sorrow. But that stability of ecosystem depends upon sympathetic stocking with sheep and cattle. If the natural balance of life on the hill is disrupted and nothing is done to put things to right, then the rolling grassy sweet Borders hill country will be lost. It may be though that the people needed to look after the hill – the hill shepherds like Bob Lamont – are all already lost.

9

Needs Go Where the Devil Drives

JEAN Retson was well over 90 and all alone. Her major concern at the time was getting the garden dug ready for her to plant the potatoes. She was reluctant to talk. But once we got going, I was enthralled. As to remembering her words, I've done what I can.

I was a Gilchrist. We had dairy cows in Dolphinton. Right from when I can remember, I was one of the gang that milked the cows; six in the morning, four in the afternoon. By hand. We got the machines later, in the 1950s; cruel things. Cow would give milk to you if you were hand-milking, the machine just sucked it out the poor beasts' tits. It's no wonder they got mastitis.

I stayed home during the war; reserved occupation – farming. So did my brother. David went out ploughing for other people. First it was all horses, then Dad and he got a tractor. A grey Ferguson. It just pulled the plough, like a horse; only quicker and stronger. My Frankie did the same at Whitmuir. Everywhere was ploughed up. That was what the government – the War Ag – made you do; there were big grants. If it were grass – you ploughed it up! If you had a tractor you went and ploughed up the neighbours too. We grew everything, not just oats and wheat. Lots of people grew potatoes. There was a big demand for potatoes – to feed the munitions workers and the miners, they said. They would be put into a big clamp. For the food reserve or something. Nobody ever came to pick them up. When they began to rot we gave them to the cattle.

But, you see, when the war started most people bought food that came in from other countries. Not much was produced at home. The stuff that came in was cheaper. Farming here was useless; most of the arable fields had tumbled back into rough grass. Then with the war and the submarines and that, we had to grow our own food; and quick about it. When Hitler found he couldn't invade us, he was minded to starve us.

Frankie and I were married in 1947, I was 27. Frank's parents moved out and went to Crawfordjohn to make room for us. We had a good herd at Whitmuir, fully 20 cows chained up in pairs for milking in the byre. All winter they were chained. They went mad when they were turned out into the fields in the spring. We never had a holiday – couldn't, there were cows to milk. The Ayrshires made no money in the 1960s. There was no money in milk. Frankie got fed up with it, he went to Wales one weekend and came back with a Hereford bull. We crossed all the cows, gave up the dairy and did beef instead. Frankie loved the Herefords and went into pedigree breeding. He was good at it. He got to sell prize bulls – good money. We had sheep too, by the 1970s the whole farm was back to grass again. The butcher came from West Linton to kill the house-pig for us. We always had a pig or two out the back. The pig ate our scraps and we ate the pig.

We bought Whim Farm up the road in the 1970s. We borrowed for the mortgage. There used to be milking cows at Whim. The man had a monkey; lived in the byre and stole the milk out of the churns. But they were gone. The place was falling down when we got there.

We went into ram breeding. I wrote out all the pedigrees. Frankie bred Suffolks to start with, then went for the Texels when they first came over from Holland. He won prizes for his Texels, and sold them for big money. He and John Boag from Blairburn next-door did it together. They both got big money for pedigree rams. It was all about how they were fed and how they looked. John was a great trimmer – made the rams look really good. They took them up to the Perth Sales and down to Bluith Wells as well as Kelso, here. With the Texel he was ahead of the game. He had the rams that everybody wanted – Texels, with the big bums and the lean meat.

Funny thing, pedigree breeding. We made no money from farming, but we made enough to live off selling bulls and tups at fancy prices to other farmers, who, like us, would be making no money out of farming. How does that work?

By the time of Frank Retson's death in 2000, Hereford bulls were no longer of interest, being outclassed by continental breeds, the reputation of the Suffolk sheep was at rock bottom, being no competition for the Texel, and as for the Texel itself, well, everybody was breeding them. But, for a score or more years, Frank Retson was a big name in Borders stock breeding. The fashion these days is for the Beltex, like the ones bred at Wester Deans – but likely, says Jean, that'll not last …

10

Stubbornness Has its Reward

*W*ILLIE *Rose of La Mancha left a thriving heavy plant hire business in the care of his sons. There are now two businesses. But that is another story. Willie's story itself is one, so it was said, of bone-headed determination. But first that name; La Mancha – hardly Borders Scots! The village is now called what the big house is called. It used to be Grange. The Grange Estate was a goodly part of the landholding in the valley from Romanno to Leadburn before the field enclosures and the making of the farms in the late 1700s. The big house at Grange was knocked down by the Cochranes who bought the estate from proceeds won in the Mediterranean where Lord Dundonald was admiral of the fleet. While at sea, Cochrane kept his family safe on the Spanish mainland, bringing back with him to Scotland the name of La Mancha.*

The new house, La Mancha house, and the Grange was sold to the Mackintosh family in 1832. For the next 100 years the family, together with the descendants of the Montgomery family at Macbie next-door, improved and developed traditional rural crafts and industries; clay pipes, tiles, pots and farm troughs, lead mining, iron ore mining, coal mining, limestone extraction, lime-kiln firing, stone quarrying.

These industries employed many, created a busy populous valley, made miscarriages commonplace among working women and permanent physical injury among the men an unremarkable expectation. Life-expectancy of a male was reduced to 32 years, while the children were brutalised into heavy manual labour.

William Wallace Miller Rose and his brother farmed the small holding at Bogsbank, West Linton. In 1940 they took a leap up the ladder and purchased the 400 odd acres of Lower Grange, which lay below the village, and with it the dozen acres in Upper Grange, which had a filling station and little motor workshop on it.

At Grange, the Roses grew 20 acres of potatoes every year, contracted to the War Ag for feeding the population being starved into submission by the German U-boat Atlantic blockade. Every year they were meticulously stored for overwinter keep. Every year they were paid for. Every year nobody came to pick them up. Every year they were wasted.

Willie developed a hankering after the big farm machinery which was beginning to come in. He spent more time on the filling station and in the garage than ever he did on the farm down in the bottom. By 1950 the love affair with farming was over (if it ever had begun). His wife Hilda, a Land Girl from Yorkshire, wanted their family out of the cold damp of Grange farmhouse. The answer was to be assisted passage (£10, children free) to Australia, which at that time was crying out for immigrants. The farm was put up for sale in two plots; farm and filling station, and quickly sold.

All was set, save for the machinations of the purchaser of the little filling station who determined that the best time to pay Willie his dues would be after he had embarked upon the boat when he would be able to do nothing about not getting monies owed! Willie got wind of this. The money was not paid and the family did not get on the boat.

Instead, his wife and the kids were dispatched to Yorkshire while Willie set a caravan behind the petrol pumps, put up a shed, bought a batch of 300 weaner pigs to fatten off potatoes, and began a new life. In the event, Hilda had all the children down in Yorkshire. She said that was so they would be eligible for the Yorkshire cricket team.

Willie purchased a big tractor with a heavy-duty fore-end loader, the only one in the district. He loaded lorries with it, at stone quarries, at Cowan's paper mills, anywhere. He loaded sand and gravel, roadstone, ash. It was Willie who knocked down the big house at Macbiehill and made the farm roads with the debris.

A house was built at the back of the plot out of concrete panels and the family called back. There was so much haulage and lifting work that Willie needed another heavy tractor, and another. He was into the plant hire business – Wm W.M. Rose & Sons.

Willie Rose had got lucky, or astute, with his timing. There were lots of farms and small rural enterprises that, while able to afford a pair of working horses kept in steady employment through the year, were now unable to afford for themselves the expensive specialised machines that were often used only seasonally. Contracting-out and plant-hire were emerging and highly lucrative businesses. Just up the road, Willie Burns at Blaircochrane did well for many years out of buying and selling second-hand tractors. Well enough to buy another two farms.

Seems maybe there is more money to be made out of farmers than out of farming!

11

Struggling to Survive

*F*OR *the 50 years that spanned the middle of the 1900s, every proper dairy farmer wore blue dungarees on his body and leather hob-nail boots on his feet. The dungarees had a pocket across the chest bib, another one in the back and two on either side. In one or other of those pockets every proper farmer carried a little pocketbook and half a pencil to go with it. Every proper livestock feed company produced one of these little diaries. They had soft card front and back, lined pages (four days to the page) and really useful information in the front – like conversion tables, feed rationing and fertiliser application recommendations, and lists of the company's products (of course). They were handed out by salesmen every Christmas in recognition of a loyal customer. As the notebook was often pulled from its dungaree pocket in front of others – in the farm supply shop, at the market, in the bank – it was important that the diary was a good* *one from a respected company. Farm offices were littered with them; every year of a long farming life faithfully recorded in tongue-licked black lead on grubby paper with pale blue lines.*

Robert Lambie of Auchendinny Mains was no great writer of prose; his entries were sparse. Other's entries were more fulsome, maybe even up to as many as a dozen words! But behind the scant words were, in those thumbed pages, the stories of family dairy farming. Recorded here are entries from diaries that could have been any of thousands up and down the country in the post-World War II era. The story however is Robert's. His diary spanned the years of the second farming revolution – from the 1940s through to the 1970s.

1942
April 5. Statement for Bank re surety.
Assets: Sheep £750, cattle £845, horses £290.
Sales: Sheep £601, cattle £273, milk £539, crops
    £924.
Margin over running costs: £1,052.

At the war's beginning, £100 was worth about £2,000 in today's money. Robert's 'family dairy farm' at Auchendinny Mains was therefore a full-on business, one that was run with his father and helped by his young sons and daughter all.

The farmhouse was substantial, with separate rooms for children. There was a clean sitting-room as well as a clarty kitchen – made dirty with boots coming in from the yard. The kitchen was also the main living and eating room. The buildings at Auchendinny Mains were made of stone and fit for purpose. The byres held 45 dairy cows – a herd of good size – still being hand-milked by the dairymaids.

One of the previous generation of Lambie family dairymaids was Agnes Mitchell, who sadly lost her mind. 'Big Aggie' was well-known by the local young lads, who would tease her. She lived rough in the Pomathorn Woods, where firewood was plentiful. She sustained herself with what she could scrounge and washed it down with methylated spirits. The local poultry-keepers did not seem to mind a few eggs and even the odd bird 'going astray'. The Lambie family knew her story, although few others seemed aware.

Big Aggie was not big, it was the multiple layers of clothes that she wore. In the winter, if it got too cold for her, she would sneak back to the farms where she had worked to sleep in the barns. If she was found, Robert – or Andrew Lambie if it was his brother's farm at Pomathorn – would be sure to take her matches from her, before letting her bed down with the cattle. In the morning she would wash in the cow's water trough before making her way back up to the woods. Big Aggie survived, independent and contrary, through to the 1930s before being mercifully taken by pneumonia. Her life story is to be found in a war graves entry: 'War Graves. Private Thomas E. Pendrich (No 40855, 12Bt Royal Scots), husband of Agnes Mitchell, of Mount Stewart, Pomathorn, Penicuik. Killed 1917 aged 30.'

Altogether, in 1942, in the middle of the next war, Robert Lambie's farm at Auchendinny Mains gave direct employment, of one sort or another, to 12 people.

Each one of his 45 cows had a personality. These girls were not numbers, they were characters. The compliant, the belligerent, the noisy, the kickers, the greedy and the hat-racks. They were milked by people who loved and respected them; made their lives with them. Robert's cows would be three years in the making and likely a further eight as full members of the dairy – part of the family. His cows, like everybody else's, had their own names. Cows through the country had names like Gerty and Spotty (always together), Fat Ermie, Fiona, Netty, Isla, Champ (had been

Plain-Jane until she won at the local show), Blue, Brenda, Agnes, Lizzie ...

Robert Lambie was a farmer of substance, but also one that joined the team to milk his own cows early every morning and late every evening. He kept his own books and walked behind his own plough-horses. For his hobby he bred Clydesdales. Family holidays were in the caravan, parked 51 weeks of the year round the back of the stackyard (until, that is, it was blown over and smashed in a gale). It was washed off, had its tyres re-inflated and pressed into service once every year – after the hay was got in and before the corn was ripe – to be towed, with the family, down to the seaside at Pease Bay.

Robert turned the 'profit' back into Auchendinny Mains; procuring machinery (to replace the human farm workers), improving his sheep (buying expensive Suffolk rams, whose price bore no relationship to their utility), improving the dairy herd (buying expensive Ayrshire bulls, likewise), breeding the Clydesdales (which would be worthless within the same decade) and betting on boxing matches.

It has to be said, however, that into the farm costs were shovelled many items that a normal family might expect to pay out of their post-tax take-home pay. Farmers saw no shame in the legitimate practice of putting the mortgage for the family home together with that of the farm, and thereby not only gaining extremely favourable interest rates, but also putting down the family home as a farm-business running cost that could be set against income. Similarly accounted as business costs were the car, the family's use of electricity and coal, the rates, water, work clothes and so on. Food for the family table moved seamlessly and without record from farmyard to kitchen: pig and lamb meat, milk, eggs and poultry. Bigger farmers would pay – as part of the farm-staff wages – a gardener to come once a week to look after a sizeable vegetable patch and soft fruit cages. Certainly the gardener (and the farm workers) all benefited when crops were growing faster than they could be eaten, but mostly the flow was into the farm kitchen's cooking pots and preserving jars.

In a word, the farm house, its residents, and all its expenditures were counted as part of the business. On the other hand, for the never-ending slog that is dairy farming, there were no wages for the family! Robert all his life worked for his father for nothing, on the grounds that he would – eventually – inherit: 'Don't fret son, it'll all be yours someday.' Robert was like many farmers' sons who did not find themselves in charge of their own businesses – master of their own destiny at last – until they themselves were of senior years; by which time they had had enough and lost their health and heart anyway. It was the same for Robert's sons; though by the time they came to get it, the farm business had, like a cowped ewe, gone belly-up.

Through the 1940s and 1950s, farmers got mechanised. Typical of diary entries would be:

1945
June 10. Bought Massey-Harris Binder; red.

1952
May 25. Bought tractor. Alice-Chalmers; horrid orange.

1954
Highland show. Bought tractor. Fordson Major; big blue diesel, with belt wheel. Special show price!

1955
Autumn Fair. Sold last three Clydesdale fillies. Given away. Probably went to Holland for meat. Shame.

1958

Aug 1. New toy. Ten-foot cut combine. Massey-Fergusson. £1,880. Enough to pay 4 men for a year. Will save two. Money back in two years!

After 1945 nobody would be purchasing binders any more. Cutting and binding sheaves would begin the slow cycle of hand labour; stooking, carting, stacking. Then through the long winter months the sheaves would be taken from the stacks and into the threshing barn. By the 1960s all this was gone and farmers' harvest-time diaries were full of the affairs of their proudest possession; the combine harvester.

Before the era of the combine, if the season had been wet, the stooks from the binder would be left to lie out sodden. Finally, if the pigeons had not already got the lot, the ears would sprout green shoots and the grain would be lost. Sometimes, out of sheer frustration, the sheaves were gathered-in damp; then they would rot in a heap in the stackyard.

Robert had got fed-up with seeing good crops wasted. The combine both cut and threshed in the field as it went along. The grain was beaten from the straw by a drum revolving at high speed, which caused it to emit a banshee howl that could be heard for miles (Robert loved that). The combine delivered straw – ready for baling – out of its back, while the clean grain flowed straight into hundredweight sacks, which were bagged-off and carted into the grain store. Job done, all in a 'one-er'. Very efficient, very labour-saving.

But for Robert, the machinery and the crops were not the reason why he loved his farm. It was the dairy cows that were the axis of Auchendinny Mains and the pivot of Robert's existence. He just loved looking after dairy cows. They were just big soft silly girls. The combine harvester had a dramatic effect upon farmer's diaries:

1959

Aug 20. Engineer out to get combine going.
   Seized up. I see too much of that mechanic.
Sept 5. Half-a-day lost while the sun shines.
   Combine drum jammed with straw – drive belt broke.
Sept 28. Raining rain rain. Harvesting stopped.
Sept 29. At last got combine going. Barley finished, but got bogged down in the wheat. Horses never got bogged down.
Oct 10. Got grain dryer installed. Noisy beast but works.

Oct 21. Even with the wet, the yields are up
  again. Seed merchant says it's the new varieties
  he sells. More like it's all that fertiliser from
  ICI that goes on. I'll ask the College man at
  tonight's meeting.

1965
29 Sept. Gave Henry his cards. He'll get work at
  Cowan's paper mills.

Robert's wages bill hardly changed in 20 years,
being around £2,500. But the number of people
employed for that money fell from 12 to four.
Following the 1947 Act, minimum weekly wages
for farm labourers was set by annual government
review. Over that same period they doubled.
But it was more than that. Whatever the wages
were, machines were cheaper. As profit margins
reduced, Robert had only one choice; increase
the size of his business or decrease the size of his
outgoings. Labour costs were the major expense.

Robert shared his way of life with those that
worked for him – *with* him. Farmers like Robert
were not sending strangers down the road, they
were dismissing people they went to school with;
rivals for girlfriends, members of the teenage
gang that racketed about the countryside climbing

trees, shooting, fishing, bird's nesting. Getting rid of their workers gave all the farmers trying to make a go of difficult times an ache in their gut – an ache that would often only go away with a glass or two of whisky.

The next generation of country folk would be opting out of living the rural idyll of their own accord if they had not already been pushed out. They would become suburban folk with sensible working hours and cash to spend on proper holidays.

1952
April 5. Statement for Bank.
Assets: Sheep £720, cattle £4,035, horses £20.
Sales: Sheep £1,605, cattle £627, milk £5,513.

Between 1942 and 1952 Robert expanded his business, and every part of it increased in profitability. He more than doubled the size of his dairy herd, and each cow in it was giving more milk; being better fed and more healthy.

Clement Attlee's Labour government and Tom Williams' 1947 Agriculture Act had created a perfect sunny climate for Britain's farmers. The purpose of the Act was to ensure that a hungry British Nation got fed from its own farms.

Money was poured into agricultural research and development over the next 30 years bringing a golden age of new farming technologies; machines for tilling the land and milking the cows, nutritionally balanced stock feeds, selectively bred crops and livestock, new industrial fertilisers, magical chemical weed and pest control.

Incentives were granted to farmers to implement these innovations. There were capital subsidies for buildings, land drainage and the like. Farmers were given free advice to help them follow-through the new ideas and focus their attention on intensifying their efforts. Yields of grains and root crops, and production from egg-laying hens, pork-producing pigs, broiler chickens and milk-giving cows all doubled in less than a decade.

And last, just in case market prices might fall in the face of increased supply, the government guaranteed that they would pay farmers directly any negative differences that might arise. The unbelievers called it 'feather-bedding for farmers'. The believers pointed to a nation that now had the luxury of a plentiful supply of cheap (i.e., subsidised) food that every member of the population could afford.

For Robert Lambie, all of this activity was centred upon his dairy cows and their milk. He automated with machines to milk the cows, hugely reducing his labour costs. He drained the pastures, and spread high-nitrogen fertilisers upon them – thereby dramatically increasing yields of grazed grass and hay. He went pedigree with his Ayrshires, and got his girls tuberculin-tested – guaranteed free of tuberculosis.

By the early 1960s Robert was milking more than 70 cows every day at Auchendinny Mains, which was, at that time, a substantial herd. He had a milk-bottling plant put in; selling his milk in local shops and on his own milk-round. He bred his own herd replacements, selling any spare heifers at good prices. He was a well-renowned and successful farming leader. He had around 200 dairy Ayrshire cattle on his farm, and he named and loved them all.

From farm milk sales and herd numbers jotted into his little notebooks, it can be calculated that in 1946 the yield of milk per cow per year was around 450 gallons. A decade later, Robert's cows were yielding 1,000 gallons each per annum.

Robert was doing more than his fair bit to feed a hungry British population.

MILK BOOK – Auchendinny Mains
1946 – 18,918 gallons
1947 – 14,771 gallons
1948 – 21,701 gallons
1950 – 34,292 gallons
1960 – 39,498 gallons
1965 – 47,283 gallons

There was something special between Robert and his cows. Lots of dairy farmers were the same. This is difficult to grasp, as milking a herd of cows is a drudge. Robert had to put up with a 365-day commitment, the unceasing toil of the twice-daily milking, and a 5am start to every day. Such irrationality has only one explanation; infatuated adoration.

Robert Lambie's dog-eared diaries had, apart from their characteristic economy of words, two notable omissions. The first was that the war was never mentioned. Robert himself and his family were of course in 'reserved occupations'; his job was to provide local food while Germany's U-boats sank British supply ships in the North Atlantic. Perhaps it was because the farm notebooks were just that … farm. And the farm was not at war. The second omission was any mention of foot and mouth disease. But he got it alright. I can, however, speak for him because of my own experiences with that most dreaded of all animal plagues. I know that he was infected from the evidence in his milk book, and reference to trips to Ayrshire to buy cows in milk. The year was 1947/8.

In 1948 Robert Lambie's farm turned in not the usual income of £1,000–2,000 to which the family had become accustomed, but a frank loss of £167.

The consequences of an outbreak of foot and mouth disease were/are horrific. There was only one outcome from getting foot and mouth. All the stock on the farm were summarily shot. Slaughtering-out was (and remains) the only viable solution. Some compensation for the value of the animals was paid by the government, but that did nothing for loss of income nor for the trauma of seeing ones livestock – in the case of Robert's dairy cows, one's friends – destroyed in front of your eyes and your livelihood taken away from you.

The 1947/8 outbreak in the Scottish Borders was notorious for its virulence. The dread of infection dominated everything. As the disease raged through the countryside, threatened farms shut up shop and cowered in fear. Robert locked up his farm gates. He allowed nobody out of or into the farm. The foot and mouth virus travelled on the wind, but most potently it was carried into farms by people and other animals. All contact by the Lambie family with other human beings was kept at a tolerance level of zero.

The Lambies were justifiably terrified. Robert, his brother and their father were staring in the face the ruination of their life's work.

The first farm to be hit in the vicinity was a

quarter-mile away from Auchendinny Mains. The virus was carried there, it was said, by contaminated bones being dumped from the road into a nearby wood.

Robert's harsh isolation tactics were not enough. He was not to be spared the scourge. Whereas in the weeks before, the isolation of his farm and family was his own choice, now *he* was the pariah. *He* was contaminated and to be feared by all around him; neighbour, friend, relative. And his children the same; outcast and isolated.

The police called around the instant the veterinarians confirmed his flock (for it was the roaming sheep that caught it first) had the disease. Then the knacker-men came, mob-handed and armed with their killing pistols.

The family was shut into the house while Robert and his stockmen organised the destruction of all that they had and loved: the Ayrshire milking cows who had served at the farm all their lives; the home-bred heifers – 15 of them heavily pregnant; the new-born calves, doe-eyed and innocent; the sheep – the whole flock and the little lambs went; the yard pigs. Robert saw them all go, shot in front of him. He was a long time weeping; he never really got over it. The farm fell quiet with death.

In some places in the Borders the carcasses were burnt, black smoke from the funeral pyres swirling high up into the air. Many folk were convinced that it was the burnings that spread the disease, uncombusted contaminated skin being swept aloft in the hot up-draught. So Robert's animals were piled – bellies bloated tight, stiff legs stuck out straight – into deep pits and covered with lime. The unnatural contour of the land that is evidence of those burial pits can still be seen in the field adjacent to the old steading to this day.

After the carnage, every building, corner, nook and cranny on the farm was disinfected. Scrubbed clean; at least there was something for the farm staff to be doing. Some farmers burnt their hay and straw stocks as well, lest the virus was hiding there, ready to reinfect. Robert kept his feed. He had to get back up and going again. He was running a dairy business, he needed milk to sell and cows to milk.

Robert Lambie knew what to do. And in this he was helped by the whole family. Farmers took what life doled up to them – you just got on with it. He could restock after six weeks. He would need to buy fresh-calved cows in full milk. He would go to Ayrshire and get the best; turn adversity to advantage. Or at least make something out of the mess.

The family was resilient. Robert planned. He had plenty of spare time on his hands and thinking about the immediate future was a sight preferable to thinking about the immediate past. He planned a new dairy herd. He travelled to the best pedigree farms and markets in the west of Scotland. He spent. He secured a revitalised livelihood for himself. The new replacement cows came in. First a score or so, to be carefully watched for any sign of reinfection. The milk lorry was got on the move again. He had fought his way – and his farm – back into life.

No sooner this, than the news came through that his neighbour had restocked a mite too soon; they were reinfected. That farm was slaughtered out for a second time. Was it all going to happen over again? The black cloud descended once more upon Robert Lambie. He shut up the premises and hunkered down: waiting. This time it paid off – he was reprieved. But the movement order was back on, which frustrated his plans for restocking.

It would be fully a year before Robert and Auchendinny Mains was returned to normal farming – normal trading. The losses to the business had been huge. But the losses in animal and in human suffering were far greater. And in the human case, the effects were permanent.

Robert Lambie never really got over it, although he seemed to – he was a farmer, he shared his feelings sparingly. Auchendinny Mains got over it. It had to, it was a dairy farm.

As we have seen, Robert Lambie of Auchendinny Mains farmed through the second farming revolution, spanning the times of hunger in the 1940s through to the times of prosperity and food plenty in the 1960s and 1970s. By the 1980s, farming's success caught up with itself. Europe was awash with milk and butter (and most other farm products as well).

This, however, was not to trouble Robert Lambie. The war years had taken its toll. The farm lost its verve. Profits dwindled. In 1967, Robert's diary noted a farm income of £685. He was depressed and dispirited. Milk was losing money. To stay viable and support the family, Robert would have had to get his herd size doubled up again to at least 200 milking cows. Besides, the bottling plant was finished, falling foul of the ever-more stringent food and hygiene regulations. Robert was not thriving. He was not enthusiastic about meeting the new challenges of large-scale intensive-farming. Neither were his sons. They had seen enough. They wanted a better sort of life for themselves, their wives and their kids.

Robert Lambie sold-up Auchendinny Mains. The steading went to a property developer, the land to a neighbour. The last entry pencilled in his little notebook was:

1969
Sept 16. Last lot of cows down the road – £3,620.

12

Women's Work

*T*HE *old lady – Miss Williamson – cracked in before I had time to switch on the microphone and put my iPod out onto the coffee table. I knew better than to fumble about. So I just sat back in the well-cushioned sofa and listened while sipping black Earl Grey tea from a spotlessly clean piece of Spode. It was at Willie's insistence that I had sought her out. Willie Chapman lived on a little 30-acre croft next door to where the Williamsons had been at Wester Deans. So I should, I suppose, start with him.*

My father, John, was thrown out of Leadburn Mains; 1953 that was. The landlord wanted to sell it up. Dad couldn't buy it. He was just a tenant; it wasn't much of a living. The owner must have felt guilty about throwing Dad out. There was this little shepherd's cottage here at Redford Hill. So a few acres around the cottage was fenced off and Dad were given it for what he could afford (nothing probably). Anyway, I was a just a little lad, and across from us were these two spinster ladies.

They fascinated me. They were so different – special. They ran this great big farm. When I say ran it, I mean they farmed it. They lambed the sheep, drove the combine-harvester, shifted the cow-muck out of the stock-yards with hand-forks. They did everything, just like any other common farm-hand. I used to go over to be with them and help. From when I was really quite young. They educated me those old girls. Bunty was the matter-of-fact one. Mary had more of the Lady about her. She liked hats. But they both wore heavy farm clothes and leather boots; most of the time they were pretty mucky. Bunty's dead. She built a cottage for herself at the farm gate and retired into it when her hands couldn't grip a shovel any more – arthritis. She didn't live long. Maybe she just got bored – retiring wouldn't have come natural to her.

What about Mary?

Still alive; gone posh. Living in a house with a turret and a garden and a lawn in Peebles. You should go and see her!

125

So I did.

The house was indeed large and turreted. Big oak gates stood open onto a wide drive and big lawn. The outlook over the town was such that whoever built the house there had obviously chosen well. There was a grand porch with coat-rack-cum-walking-stick stand. I looked for the expected Jaques croquet set. Mary, I was given to understand, had all of the ground floor of this grand mansion to herself – there was another flat above. The hall and the rooms were generous. The lounge, into which I was ushered had within it a blazing fire. The furniture was dark, Edwardian, not too big, rather well-chosen; there was a lovely desk that I immediately coveted. The walls were adorned with scenes from the highlands or the fishing villages of the Fife coast like Pittenweem with its red roofs. On the side-boards were pieces of ornamental porcelain. The predominant colour seemed to be pink, but I'm sure that was more of an impression than a reality. Mary was slight. She was bent to be sure – she must have been about 90 – but she had dignity and bearing; dressed in smart cottons such as would befit the lady of any grand manor. She reminded me of my own maiden aunt who had never married – her man was lost in the Great War – and had taught at Cheltenham Ladies College. Mary made a great cup of tea served with her own shortbreads.

If Mary was now living her life as she had always wished it to be, she must have had a lifetime of deepest misery on the farm at Wester Deans.

My father brought the family across from Moffat. Mother and us three girls.

When was that? All girls?

Maybe mother lost some, she never said. Father wanted boys to follow him into farming. Maybe he got them. He took the tenancy. The owner lived in Australia or somewhere. Never saw him, never corresponded. He didn't care. We never got any landlord's farm improvements. All the capital had to come from us, even though the landlord should have done it. We only saw the estate agents. Up-market they were. They just got the rent from us and left us alone; 1933 …

We couldn't live in the farm house. Its insides were all burned out. It had been a lovely place – big rooms, patterned plaster-worked ceilings They said that the lord at the Whim, Stanhope, had set it up for a gentleman – was it a Mr Stevenson? – that was to be the biggest dairy farmer in Peeblesshire. To supply the gentry in Edinburgh with milk and cheese and butter. Sometime, maybe 100 years ago, afore I was born anyways, there was a big dairy herd, a Mr Mark was the farmer by then. The Marks had lots of places all around. A daughter went to Walltower farm in Howgate, married a Pow, most of them seemingly ended up in the lunatic asylum.

We were told that before us, the place had been farmed by the Leans. Two families, all related all over the place, it was like a village of just one family. There were two or three brothers, a sister, at least one wife and countless children. The shepherd's bit at Blinkbonny – Blinkbonny was part of the farm – had half of them and the farmhouse up at the dairy steading had the rest. There'd be 20 of them or more. They had all pushed off to Wales we were told.

She sipped her tea, falling silent, gazing into the fire. She had pronounced 'Blinkbonny' with noticeable affection. I guessed she was there now, and who was I to interrupt. Soon, however, with a quiet sigh she rose from her straight-backed chair and put another log onto the flames.

Am I boring you? An old woman's memories. What do you want with them? What you going to do with my cast-offs?

No, please go on. It's perfect. People have no idea now about farming, about how it was. I want to tell them. Was it the Leans who destroyed the farmhouse?

Heavens no! That was the man that came down from Orkney.

Orkney?

Aye, I don't think the Edinburgh estate agents even knew where Orkney was. Anyway, he takes up the tenancy after the Leans left. His housekeeper up there was coming down to join him at Wester Deans as his proper wife. He gets settled in, but finds his new wife-to-be has been two-timing him for many a long year as she was already married. There's a big bust up by all accounts, the house-keeper pushes off to goodness-knows-where, and *himself* gets depressed, goes into a rage, sets fire to the farmhouse and departs – presumably back to Orkney, but nothing was heard.

When Father arrived with his cattle and sheep and everything from Moffat, he had been told nothing of this. The house we had come to live in was uninhabitable. Mother took us all off down to the shepherd's cottage at Blinkbonny and set up home. It was falling apart, but better than the farmhouse. We had a good childhood there. It took the estate agent people years to do up the farmhouse, and then it was a poor, cheap, job. At the beginning father found it hard to make a go of it. As I remember we lived mostly off rabbit. There were plenty of rabbits. Daddy would catch them in snares and bring them home for the pot. We all got fed up of rabbit. Don't ask me to eat rabbit now!

Our Mother, Isabella – Dad called her Belle – died in 1949. Father was stuck with three girls. He treated us like we were boys and we were all set to work. Slave labour it was. We were called 'the Williamson girls': Agnes, that's Nancy; Alice, that's Bunty; and me. Agnes got fed up with it and the next year ran off with the boy from Ruddenleys over the hill. They ran away to foreign parts; to Eddleston two-mile down the valley.

Mary smiled at her own quip.

Maybe it would have been better if Bunty and I had found two more boys to get *us* into trouble so we could have gone too. But we didn't. Father set up a company. He always wanted it to be 'John Williamson & Son'. But he had to make do with second best, setting up with us girls as the partners – 'John Williamson and Company'. That same year – 1950 – we made the landlord an offer and bought him out. Wester Deans was ours now. Ours to pay the mortgage on. Father died in 1960. After that Bunty and I just got on with it. What else was there for us to do? Farming was what we knew. We were alone with the farm. We just got on with it.

* * * * *

They had 'gotten on with it' for the rest of their working lives; another heart-breaking, body-breaking, 26 years.

I went to visit Wester Deans soon after my visit to Mary, being shown around by the manager.

The farm was being managed by a competently experienced college graduate who would other-wise have been farming on his own account in Dumfries. His misfortune was that his inheritance (a large upland sheep farm) had been sold out from under him and put down to a vast expanse of forest plantation Sitka spruce.

By the house was a big five-yard square 'Milkhouse'. There were stone benches a yard wide all around – presumably originally for cheeses and butter. In Bunty and Mary's time, the dairy had been used for the paraffin generator that made electricity for the farmhouse.

In the old steading, standings for four working Clydesdales were still evident. With four horses, the girls would have been ploughing 60 acres a year. Bunty's first tractor would have come as great relief to her, likely not long before her father died.

The brochure for the farm when the sisters put Wester Deans up for sale in 1986 was illuminating. Bunty and Mary had been running, by themselves (with a bit of help from young Willie in his school holidays), a 400-acre farm with a flock of 500 lowland ewes to husband, lamb and shear (a full shepherd's worth), 70-odd suckler cows with their calves (a full stockman's worth). Of the ploughable acres, it seems the two women grew more than 50 acres of barley, a similar acreage of other crops, and made enough hay for winter feeding (at least a full tractor driver's worth). On top of that they did their own contracting, driving their own self-propelled Massey-Harris combined harvester and a baler. Then there was all the office work (at least a day's worth a week), a business to run (another day) and a home to keep. The amount of toil that those two women got through is almost impossible for us to comprehend today. Their father must have trained them well.

13

A Good Doing Over

ABOVE the valley stands a fine long hill ridge. Upon those hills not that long ago were grazing lands for cattle and sheep. There were steadings with low stone buildings and cottages. There were clumps of trees; oak, birch, beech. It was a haven – not just for wildlife, but for upland farming life too.

The man who had accosted me in the middle of the street had lived in the village of West Linton all his long life, as had his family's previous generations before him. He had parked his mobility scooter next to a weedy flowerbed. He was gazing up at the flank of the ridge of hills that look north and west down to the Lyne. Up there was Roger's Craig, the Kelly Heads, Deans, Grassfield, Ruddenleys, Fingland, Cloich.

I love those hills. We used to spend hours up there. With our slingshots – catapults to you

– getting rabbits. Or grouse, or pinching plover's eggs, or whatever – we were just country boys. A couple of posh girls would gallop their ponies up there. We liked watching them. It's easy to fancy a girl on a horse – 'specially if she's posh.

It were a happy place. Look at it now – it's been raped till its dead. It's covered in trees. I hate those trees. I just *hate* those trees. They're the wrong green. They're like a field of wheat, not like a forest or a wood. They're solid. There's no space.

He grasped me by the arm and steered me to a seat by the council rose-beds. 'You write stuff, don't you? Well, tell those bloody townies they are killing our country with their bloody trees. Tell them to stop.'

Look, when we went up there it was all grass. Damn it, the place is *called* Grassfield. There was a farm up there – Grassfield Farm. Sheep, cattle. The cottage, a byre, sheep pens. Out along toward Flemington there were big patches of bilberry; we picked them to bring home. Few of them got home mind – we ate them on the way. There were lots of grouse, black grouse, and partridge. And there were little birds that nested in the grass tussocks. Meadow pipits, skylarks – mind when you've ever seen one of them now? They've gone. There's nowhere for the animals to live in a Christmas-tree forest.

The sheep. They all had their own bit up there. They didn't go where some farmer put them. They went where their own sheep brain told them to go. The shepherd was just there to keep them right. His flock had its own hirsel, the flock's own space – their hill. The hill and the hirsel and the sheep flock are all together. Nobody has a right to separate them. Then on the hirsel each sheep family had a heft; their own ground. Used to be common grazing years back. Nobody owned it. It belonged to everybody; loved it. It was good. Now? Now it's owned. So it's owned and because it's owned it's just property and it's sold. Sold and planted with trees because they buggers in London get a big tax break for making a stinking American forest. Where we used to play. And the farms.

You know what they're going to do? They'll cut the lot down. It'll be a bomb site. Then they'll see the trees have sent the soil all to hell, so there's nothing else'll grow but more Christmas trees. So there'll be more trees. They'll make out its all natural and environmental by planting a few larch round the edges. More tax breaks. And likely, you know what, the bastards who live in England don't even know where's Roger's Craig. What'll they do next? Put a whole line of bloody wind turbines there – right along the skyline. Who's going to want to go up there now? No one. Not man. Not beast.

He got up pushing on his stick. As he stumped off he turned 'Aye, like I may be sorry. I was forgetting like. You're *English aren't you?'*

14

Jane Bartie and Doocot

*W*ALLACE and Nancy Dow's farm was made just the right size for a family; when they took it over that was, early in the 1980s. Upland farming is never kindly, but in the past 50 years, Nancy and Wallace will tell you, it has been particularly unkind.

Their farm, running down to the Lyne from Drum Maw and the hills of Whiteside, is upland livestock country through and through. Since 1984 when Wallace and Nancy moved in, the fortunes of upland livestock farmers have drifted inexorably downwards. Now, at 275 acres, the place is getting too small to support a family. The Dows are typical of hundreds of small upland farmers who have battled it out over a generation. Not just with the land and the weather and the sheep and the cattle, but also with the market and the politics. The only solution to high costs and low returns has seemed to be the redoubling of one's own hard physical labour. Yet these same farmers, all across Scotland are still there; undefeated. They have not won. Theirs is a battle that can never be won. But just to remain, not yet beat, is presumed good enough. This is what makes Wallace and his likes angry. There is so little to show for a life genuinely spent. There should have been more than just survival for a generation of hard grind, year-on-year.

Scotland's upland farmers have a feeling that the world has let them down. It is not that they do not love their farms – they do. They love their cattle and their sheep. They love the unkempt braes of the hills with their wild upland flowers in the pasture. They love the native broad-leaved trees around the big field boundaries, and the ancient drains and burns rumbling as they carry the hill water down between their stones. They have risen to the challenge of producing beef, lamb and wool from land that can have no other purpose. Beef and lamb that cost more to produce than the market will pay for, and wool that is worth less than the cost of shearing it off. The annual shearing used to be a happy, convivial – if sweaty – time; and a wee bit of profit in the wool-sack at the end of the day. Now it is a thankless back-aching chore. The only purpose being to give the ewes relief from heat and flies.

That is the lot of all farmers of marginal-land. Wallace fears for the future of his, and all the other upland farms; the ones that need to be the right size to be worked by a family team and support that family for their work. Their *time surely must not be over. Wallace and Nancy Dow do not wish to be the last of the line.*

But the story of Doocot Farm is not about Wallace, it is about his mother. It was told me by Wallace and Nancy in their kitchen. There was a cup of tea, a scone and a natter about how things used to be. Nancy's kitchen, like all farm kitchens, was cooking place, eating room, library, office, living room with settee, television room, warm spot by the fire, bar. I was taking notes at the dining table – or was it the desk? – trying to listen while avoiding looking at the small television perched up on a shelf talking to itself. Before Wallace launched into the dire state of Scottish Borders farming, Nancy thought fit to first explain the name.

There is not any dovecote at Doocot Farm; not a semblance of one to be seen. Nor any memory of one. There would have been one, a large one, right next door, however, at the big house – Romanno House. Doocot Farm got its name from a famous fight that went on there between two bands of Romany gypsies. They had had, it seems, a good day somewhere (perhaps Edinburgh way), and were on their way to their next gathering when an alcohol-fuelled altercation broke out between rival families who had been until that moment good friends. It is not known what the bitterness was about, but a few people got knifed. Murder most foul had been committed. Blood spilt on innocent soil. There was a trial and hangings to follow. It all happened in 1677. The whole notorious event was recorded by the surgeon rhymester, Dr Pennecuik of New Hall. Romanno was in his ownership at the time. So, being the owner and being a rhymester he naturally put up a dovecot on the very site of the appalling events. Over the door, on the lintel, he had inscribed: 'The field of gipsie blood which here you see, A shelter to the harmless dove shall be.'

Anyway, even after all of that, the properties that Pennecuik had in Romanno were let to the Montgomerys of Macbiehill before being sold off. Of the doocot? Nothing!

Now, as to Wallace's mum. Every new resident of the village of West Linton for the last 100 years has, within a few weeks of their arrival, been inculcated with the story of the three Girl Guides who fell to their death off the high cliff that makes the dramatic ravine north of the village through which flows the river Lyne. Along the cliff top runs a narrow unfenced path, the 'cat-walk'. It is indeed dangerous, being especially slippery after it has been raining (a not uncommon occurence in Linton). The walk is popular with the village folk and their dogs. Hence the need for newcomers to be warned through the medium of what has now become a well-rehearsed urban myth.

Myth it is. Three little girls tumbled down the sheer ravine to be sure, but only two to their death. The third was Wallace's mum, who it seems, avoided the final plunge.

In the 1920s, Jane Bartie was a Brownie in the pack at West Linton. Brown Owl was the lady wife of Captain Thomson of Kaimes. A good choice as the family had a big car and a chauffeur to drive it; a most useful attribute for any Brown Owl to have. Jane was the 'cat-walk' Brownie to survive because she had leather soles to her boots; the other two had rubber. It was slippery, for it had rained the night before. Their expedition to pick springtime primroses on the steep slopes of the Lynedale Gorge above West Linton ended as the two slipped and fell to their deaths over the cliff edge to the stony river 80ft below.

Jane's father was the slaughter-man at the abattoir along Linton's Deanfoot Road. The main business of the Bartie family was however a little more refined; dairy farming at Mansfield Farm in Newton Grange. Jane found herself employed there in the early 1930s. She spent much of her day learning the trade by pushing a handcart around the miner's cottages. She doled the milk out of five-gallon churns with a ladle into the quart tin-cans held up for her by the children; some of whom would die of the tuberculosis so easily transmitted through raw milk.

Jane Bartie grew to be a tall and elegant young lady with a shrewd business brain. Within a few years she had at her disposal a grand milk float drawn by a fine horse called Tammy. It could handle many more churns of milk and a much larger milk round. Jane was in charge of a dairy business! On top of this, as expected of a woman of the time, Jane got married to Marshall Dow and had two sons. Marshall died tragically young and Jane was left with both a dairy business and a family to run. She was ambitious for her boys. The eldest wanted to go farming.

When Low-Mitchell put Doocot up for sale in 1962; Jane bought it. Remarkably, the Low-Mitchells had a small herd of Jersey milking cows there and a little bottling plant. The herd could only have broken even on Doocot's difficult country by selling the yellow milk at a premium. Which they did, sending it to Edinburgh town for consumption by the lawyers and gentry there – a rather different sort of market to the miners of Newton Grange. In retrospect, however, it became apparent that even with luxury prices on the milk, Doocot Farm lost money.

Everything was wrong. Through the 1960s, incomes from dairying went through the floor. The herd was too small. The country was not suited to Jersey cows – it was upland beef and sheep land.

On top of all of that, Jane's eldest son decided he was no longer enamoured with the unremitting toil of milking cows twice daily. What took his fancy was driving big haulage lorries. He did not like the idea of farming at Doocot anymore. When he went in 1970, Jane sold off the dairy herd for a pittance and the farm was restocked with beef cattle and Frank Retson's Hereford bulls to run with them. The whole show was managed from Newton Grange, 20 miles distant, with a herdsman put in charge at Romanno.

This too was doomed. The cows fell barren and poor grazing management caused the bottom fields to fill with noxious yellow ragwort weeds. The farm had become, in a word, neglected. Into this disaster zone Jane put her second son – Wallace. Wallace and Nancy got stuck in with the ever-ready assumption that hard graft would be enough to get the family out of a deep financial hole. The ragwort was pulled, the lower fields ploughed up and good pasture grass resown. Modern bulls were bought for the beef herd and a big flock of up-to-date Texels was established at Doocot. Wallace did everything right. He slaved away doing things right for the next 30 years.

The story of Doocot Farm has been no different to those of the farms either side. One of these is fine for now, in the safekeeping of the family of a gentleman lawyer from Edinburgh. The other, like Doocot, is waiting for changed times. Sandy's story is not so unusual; farming life is full of this world's Sandy Andersons.

Sandy farms the slopes of Penria hill next door to Wallace. The heights of Drum Maw loom above, and Sandy's place is classic livestock grassland – beef and lamb – country. Sandy farms because he knows nothing else but to work the place he was born and bred on.

The steading was built by John Noble in the mid/late 1700s. At 160 acres, it is smaller even than Doocot, but the house and steading are rather fine. For some reason, a market gardener and coal merchant from Gilmerton, Edinburgh, took a fancy to the place and bought it in 1926. Alexander Wright was aged 50. Noblehall is not only small, it is difficult: a typical upland challenge.

It was a somewhat surprising change in choice of lifestyle, but old Alexander had just got himself married. Alexander farmed there until he died in 1972, aged 96. There were two daughters, Agnes and Beatrice. Agnes (Nancy) was born in 1927, she married in 1955 and little Alexander (Sandy) was born in 1958.

Sandy was a clever kid, doing exceptionally well at Peebles High School with a bright university future ahead of him, when disaster struck the family twice-over – the sudden death of his father in 1969 and his grandfather three years later. Sandy left school immediately to come home to help his mother run the farm. He just got stuck in. He is still at it; suckler cows and a flock of sheep. But there is not sufficient at Noblehall for a family to live off – nothing like sufficient. Sandy's greatest wish is for some capital to invest; in the house, the steading, the field drainage. If it comes at all, it won't be generated from the farm's income!

Noblehall was once farmed by Montgomery of Macbiehill; the one who as Lord Chief Baron of the Exchequer drove the farming revolution by pouring vast amounts of investment capital into Newlands Vale in the late 1700s, where these stories started. Scotland could do with the likes of him again maybe!

15

Farewell to Borders Farming

THE Border country has many borders; political, geographic, social. Now it has become borderline for farming too – marginal in its capacity to produce food.

But in the 1700s, Newlands valley was productive enough – Major-General William Roy's mapping in the late 1740s shows clearly how much of it was cultivated. However, by the late 1700s the run-rig tenancy system was failing the cottars. The new way of farming was a triumph of profit and productivity. The field enclosures, the draining, the tree planting, ditch digging, soil liming and fertilising, the investment in farmsteads. None of it came easily or cheaply, but it came good.

Much the same resulted from the second farming revolution which was pushed on by the

1947 Agriculture Act, guaranteed prices, mechanisation of farm crop and livestock production, and the implementation of research science into practice.

Presently however, with foreign food again in plentiful supply, the narrative of farming the borders seems to have returned to being a story of farming families fighting adversity.

The evidence of today is that the good times are again well and truly over – their demise triggered by Europe's overproduction of food through the 1970s and 1980s.

Much of the land of Romanno, between Howgate and Drochil, Blyth and Linton, seems now to be slumbering through lack of gainful employment. The farms that thrived in the 1950s are failing. Farming in the Borders has become a matter not of gently husbanding the land, but of letting it go. There is no money to reinvest into farming.

In the vale we already have in clear sight a pattern of extremes. Four or so farms are making a go of it. All are characterised by two adjectives – the first is 'big' and the second is 'intense'. Big sheds, big machines, intensive crops, intensive livestock.

And the other 40 that surround these oases

of frenetic activity? For them the hinterland advances inexorably upon the thinning flocks of sheep nibbling poor grass among the whins, breken and threshies.

If there is to be a third farming revolution, what might it deliver? Perhaps the future for the Borders is to have only few farms, but each of very large scale and intensity, surrounded by many acres of wilderness within which leisure-seekers may wander.